<u>**OCEAN FLOOR MYSTERIES**</u>

The Amazing Mystery of the Great

FACE

on the

Pacific Ocean Floor

<u>**OCEAN FLOOR MYSTERIES**</u>

The Amazing Mystery of the Great

FACE

on the

Pacific Ocean Floor

Lloyd Stewart Carpenter

Spiral Enterprises Publishing
Pasadena, California, U.S.A.

Published By:
Spiral Enterprises Publishing
260 S. Lake Avenue #177
Pasadena, California 91101

Second Edition
Library of Congress Catalog Card Number: 97-92205
ISBN: 0-9659627-0-9
Printed in the United States of America

Earth's Topography From Space

This computer generated illustration of Earth is part of the NASA Goddard Space Flight Center's program of Earth-science research. This view is generated from orbit and ground-based instruments. It shows the land topography and also the key elevations of major mountain ranges and details on the sea floor.

As you read this book, refer back to this global map.

The stunning outline of the weeping face on the Pacific Ocean Floor will become even more apparent.

An Important Part
of the

Middle East

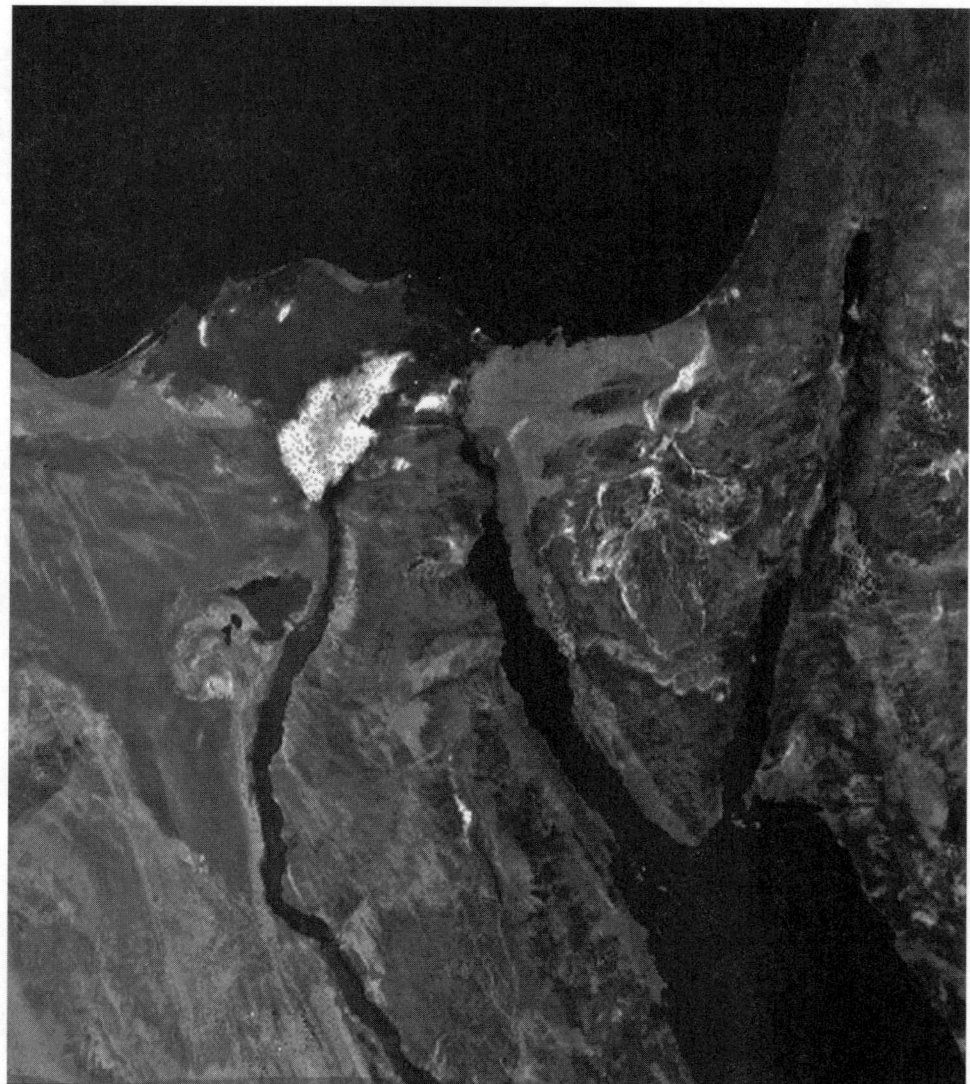

This satellite view from space shows the topography of much of the Middle East. Possibly, after viewing the evidence that can be found in this book, some readers may never look at a map of this area in the same way again. This map is from the U.S. Defense Mapping Agency and NASA. A wider view is on page 166.

TABLE OF CONTENTS

FOREWORD

By Dr. Dean S. Gilliland

When I was living in Africa, as a missionary, a famous chief died and on the selfsame night that he died a massive comet appeared in the sky. It remained there for two weeks, diminishing each day. Everyone said it was the gods honoring the dead chief. You say, "That is the kind of thing that happens in Africa." But two years ago a housewife in our California town claimed that the face of Jesus Christ had appeared on her screen door. To most, this news was either humorous or ludicrous. Yet, the number of people who went to check it out required management by the police.

Extraordinary phenomena have always had amazing fascination for people. Without warning, figures, faces and apparitions appear in unlikely places. Such occurrences seem to defy natural laws and have no rational explanation. Whether real, imagined or outright fraud, people respond to these sightings with a wide range of interest. Many simply dismiss the claims, others are curious, even awestruck, while some even feel reverence and the need to worship.

What Lloyd Carpenter lays out for us in this book is not his personal quirky vision or some passing apparition. It is not in the category of a frivolous Man-on-the-Moon or an uncanny face on Mars. The phenomena he asks us to look at are permanent beyond words, older than old, and are of super gargantuan dimensions. What we are asked to consider is in a class by itself. This report is about configurations of nature in the sea and on the land, formed by primary structures of the earth millions of years ago. It seems that, for several reasons, it is only now, in our time, that we can begin to comprehend these manifestations and reflect upon whether there is mind and message behind them.

Being in the business of theology I tend to doubt anything that we haven't heard before on the subject of God. And one could wonder why any person might imagine that he has discovered something this massive which we didn't know before. But one thing is sure. The creator God has not gone away or left human beings without means of contact with him. Rather, God is a communicator on the divine scale, so we should not be surprised when God contacts us on the grand order of creation itself.

Once the reader sees what Lloyd Carpenter has seen, responses will differ. Some may feel it is a purely subjective notion which the author has or that what is there is pure coincidence. Or you, the reader, could be led to ponder the fact that the Creator Mind is trying to tell us something. It is the latter that Carpenter calls for in his concluding section. The author has a right to ask this of us because his discovery has a history of reviews by responsible people and he, himself, is a serious thinker and scholar. But whatever the level of response, as the reader you will have to ask some big questions that have been there since the beginning of time, such as:

Is there a Mind behind nature or is it all a fortuitous accident and the consequence of natural laws? Where is the line to be drawn between pure coincidence and intelligent purpose? Is there an Eternal God who has always connected with creation and does this God have something to say to people in our day? Is there common spiritual intuition in all human beings which can lead them to similar conclusions when given the same data?

Carpenter opens up to us the possibility that we have a message from God that has been there for millions of years waiting for the right moment. Could that right moment be now? We have a new sophistication in producing maps giving us the ability to see in detail the ocean floor and the configurations of whole continents. One can imagine all of this being visible from space. It doesn't surprise us anymore to have views of the earth's surface that were never before imagined. It wouldn't have been possible to make any sense out of these formations until now because the technology was not there.

These formations are for any and all to think about, even if one prefers not to accept Carpenter's conclusions. Regardless of age, education, ethnicity or religion, no one needs to miss out on it. Carpenter's assumption is that God can do what he wants to with nature. Though the origin of these stupendous pictures reach back into prehistory, God is not bound by human time and God can use whatever means he wishes, whether seas, land, mountains, valleys, skies, etc. There is no limit to the means that God can use to say what he wants to say. This is the freedom and the power that God has.

Of course, to follow the thesis of this book raises the question as to why God would show himself in this way and at this time. After all, these natural structures have been there from the beginning. Is it because people need now, more than ever, a message from God about himself?

Is this a revelation that is made available just as we are ready to embark on the unknown of a new millennium? Even if one feels he or she cannot see anything in this material, could it not give reason to consider whether or not God is "out there" and is involved in our affairs? If God has done this to cause us to think, then the implications are almost overwhelming.

You are about to read a book that will call for some kind of reaction. If you don't come to the same conclusions that Carpenter has laid out for you, look deeply into your own consciousness, ask your own questions and see what kind of answers you get.

Dr. Dean S. Gilliland
Fuller Theological Seminary
Pasadena, California

*Dr Gilliland is a professor and also holds the Chair for Contextualized Theology and African Studies at the School of World Mission, Fuller Theological Seminary. He received his Masters of Theology (in New Testament Studies) at Princeton Theological Seminary and his Ph.D at Hartford Seminary Foundation. Before coming to Fuller, he was a missionary to Africa for more than twenty years. He is the author of several books including, *"The World Forever Our Parish"* (Bristol Books, Lexington, Ky.) and *"The Word Among Us"* (Word, Irving, Tx.). Dr. Gilliland is also an ordained elder in the California-Pacific Conference of the United Methodist Church.

****

ACKNOWLEDGMENTS

I wish to express my appreciation to the following individuals that without whose assistance this book would not have resulted in its final and complete form.

To my wife Anna Carpenter whose contribution as an artist was invaluable to the quality of the overlays used in this study. Her assistance in the areas of moral support and proofreading throughout this work has been invaluable.

To my parents, brothers and family who have supported me during this more than 15 years of research, God bless you all.

To Dr. Dean Gilliland who so generously agreed to provide the Foreword to this book. He also gave advice as my mentor with editing and additional proofreading assistance. For his help and encouragement during the final stages of this book, I will always be indebted.

I would also like to thank those professors at Pasadena City College, California State University at Los Angeles, William Carey University and Fuller Theological Seminary. Their encouragement over the years helped me to continue in an area of research that has always been a challenge.

Finally, there are many other people that I would like to thank for their assistance. I am grateful to each person that offered endorsements, counsel, research assistance, proofreading and encouragement. These people know who they are and know my appreciation.

Introduction

GOD IS A SPIRIT

To accurately describe the physical appearance of God is an impossible task. The Bible teaches that God is an omnipresent spirit (Gen. 1:2). To describe our creator as intrinsically anthropomorphic (human-like), limits everything about him. It is true that God made us in his own image (Gen. 1:26), but that is the image of love and spirit, not limited to the frailties of flesh and bone.

John 4:24
"God is a Spirit:
and they that worship him must
worship him in spirit and in truth."

Still, God is seen as having physical human characteristics at various times in scripture. Very early, we see "the Lord God walking in the garden in the cool of the day" (Gen. 3:8). This event follows the initial sin of Adam and Eve. Generations later, God wrestles with Jacob (as a man) as part of a test leading to Jacob's new name, Israel (Gen. 32:24). In the book of Revelation, the prophet John sees a vision of God sitting on a throne in Heaven. John's description of God has many elements that make the Lord seem very similar to a human form. The apostle writes that God has hands, feet, hair, eyes, mouth and a "voice like many waters." He further records that God's clothing includes, "a robe down to his feet, and a golden sash around his chest" (Rev. 1:13-16) (NIV) .

But to Moses, God appears as "fire out of the midst of a burning bush" (Ex. 3:2), and John the Baptist sees the spirit of God "descending from heaven like a dove..." (Jn. 1:32). How God may choose to represent himself is always part of how he communicates with humanity. We cannot limit God's likeness to any single event, manifestation or image. This research uses a natural image on the Pacific Ocean Floor as a metaphor to show the love and mercy of God. But God is not a map, or a man, or anything that humanity can clearly define. According to the Bible, God is an omnipotent, omnipresent, omniscient, immutable *Spirit.*

TOPOGRAPHIC MAPS REVEAL
THAT OUR PLANET
IS A
SERIES OF IMAGES

A minimum of six huge images cover the whole earth. Each is fabulous in detail, recognizable and clear. Are these images a message to humanity from God? This is something each reader must decide on his own. I have been working on various drafts and versions of this book about the face on the Pacific Ocean Floor and other images on the earth since even before I went back to college in the early 1980's. After having gathered so much evidence and associated data relating to this research, I sometimes forget what it is that impresses people most of all. It is the first and most startling discovery of all.

THERE REALLY IS A HUGE FACE ON
THE PACIFIC OCEAN FLOOR

I intend to keep refocusing on that fact throughout this volume. To do that, it is necessary to also discuss the implications of what this face might mean. The reason for this is simple. All other topographic images discovered through this map-study relate in proximity and theme to the face on the Pacific Ocean Floor. The real evidence is not only geographical but also biblical. Only four sets of images make up all of the ocean floor topography on earth. The largest is the tear-stained face of a weeping man on the Pacific Ocean Floor. This face covers almost one-half of the Earth.

HALF OF OUR PLANET IS A BIG WEEPING FACE

According to modern topographic maps, this weeping face is located at the bottom of the Pacific Ocean and encompasses all of it, including its borders. The detail of this huge face and the other images in this book are revealed through ocean floor maps, detailing the underwater mountain ranges. The scientific accuracy of modern topographic maps and satellite imagery make this unique study possible. The resulting evidence is startling. Our planet is an observable series of gargantuan pictures etched in stone on its topography.

THE FACE OF GOD?

In 1983 a half-page article appeared in the Pasadena Star News. The big headline read: "LLOYD CARPENTER SEES THE FACE OF GOD."[1] With the article was a large reproduction of the Rand McNally map of the Pacific Ocean Floor. The headline was not my idea but I had wondered about the possibility. My speeches and lectures began to include details about how this discovery might be linked to the God of the Bible.

CAREFUL CONSIDERATION

Even though I consider possible biblical links, I do so cautiously. As a minister, licensed by my denomination, I've preached many sermons over the years. But, no matter how excited I was about my maps research, I have never before preached a sermon about it. Also, I have never brought the subject up, even in passing, during a sermon.

During this period, I enjoyed teaching the Sunday morning Bible studies for the college students and young adults. Still, I never taught about my map discoveries and research or how they might relate to the Bible. My map lectures and speeches have been more scholastic than theological. Presentations have been limited mostly to college campuses. I have always felt that the church is no place to promote a work in progress, especially on Sunday morning. Of course, things do change.

There comes a time that every valid study should be published. For my study, now is that time. I admit that all my work is not finished but I also realize that it never will be. I publish as a student and not as an expert, but nevertheless, I publish.

If you disagree with any point in this book, write me via the publisher. If you'd like to correct a scriptural interpretation or scientific item, please feel free to do that. No one person can be a total expert in all the areas relating to this study, I welcome input and feedback. If I can comfortably stand corrected, then each edition of this book will be an upgrade.

Does the face on the Pacific Ocean Floor metaphorically represent the face of God? The evidence seems to support this view. Much of the early evidence supporting this idea occurred as a natural part of my college studies. As a speech communications major, I needed interesting topics for four years of weekly speech tournaments.

The professors, coaches and judges permitted me to include into competition, speeches about this map study. Many of the key details in this book have been judged by some of the best people our colleges and universities have to offer. Most of the comparisons in this book have been scrutinized by professors, scholars, judges and students. All of these people were totally serious about the quality of content for each competition or presentation involved. As a senior, I also produced a 52-minute video tape documentary about these ocean floor images. The documentary called, "Scientific Proof of the Deliberate Supernatural," became part of a national awards video documentary competition.[2]

My maps research was also mentioned in a front-page story of the Los Angeles Times. That story was picked up by the wire services and circulated throughout the world.[3] I was also the subject of an interview and story by NBC television (Channel 4) News in Los Angeles.[4] Interest in that story resulted in widespread inquiries. But even with this notoriety, I have limited my lectures and interviews. This is because until recently, there was key information still undiscovered. I still wasn't confident concerning the answer to my main research question, "What is the face on the Pacific Ocean Floor looking at and what causes this face to weep?"

WHAT A CONCEPT

Ancient philosophers declared that all things resemble in constitution and form, the human body. The Greeks taught that Delphi was the navel of the earth. Our planet was seen as a gigantic human being, twisted into the form of a ball. When Atlas was turned into stone, his beard became a forest, his bones hardened into rock and each part grew until he stood like a mountain supporting the heavens.[5] An Indian in Mexico once told me that his people have always referred to the earth's rocks as God's bones.

OCEAN IMAGES AT WAR

The images we shall be presenting are too vivid to be dismissed as coincidence. There is a message here that calls for our attention. Simply by looking at these images, we can reflect upon what this message may be. The entire scenario is not complicated when we see it piece by piece.

The dominant figure in this book is the colossal face which spans the entire Pacific Ocean Floor. This face appears to be masculine. He seems to be fixated on one thing only. What is it that causes him to weep? What is he looking at? Attacking the image on the Pacific Ocean Floor appears to be a Devil on the South Atlantic Ocean Floor. At the Devil's throat is Antarctica. Helping the Devil on the South Atlantic Ocean Floor is a dragon. We see this dragon when combining the topographic maps of the North Atlantic Ocean Floor and the Arctic Ocean Floor. On the topographic map of the Indian Ocean Floor, we see an angel (Cherubim) with a flaming sword. This angel is pulling the demon on the South Atlantic Ocean Floor, off the weeping man on the Pacific Ocean Floor.

Comprehensive coverage of this topic is impossible at this time. This discovery is still so new that there are no real experts who have knowledge of it. I am a researcher, not an expert, but like any responsible researcher, I have relied heavily on my bibliography and other people who are experts in their respective fields. This book will focus on the face on the Pacific Ocean Floor. The other ocean floor images are presented here as they relate to the face on the Pacific Ocean Floor.

Footnotes and bibliography are part of this book for the convenience of the reader and in fairness to those where specific information is retrieved. This book can only document a small fraction of what there is to be learned about the face on the Pacific Ocean Floor and other images on the earth. It will take geographers, computer experts, mathematicians, and theological scholars (working in joint effort and independently) to adequately explore this phenomena. Perhaps by studying these images and seeing how they relate to each other, we can learn more about the God of the Bible, and the age old struggle of "good versus evil."

All of the ocean floor images in this book are very interesting. But none of them come close in size, structure, detail and importance to the face on the Pacific Ocean Floor. We see the right side of a man's head. His expression displays an emotion suggesting mercy. We see by his tear-track that he is weeping. What is the focus of his attention? His face covers almost one-half of this Earth. **Whatever the face on the Pacific Ocean Floor is looking at, it must be the most important thing in the world!**

Lloyd Stewart Carpenter

Chapter 1

PLANET OCEAN

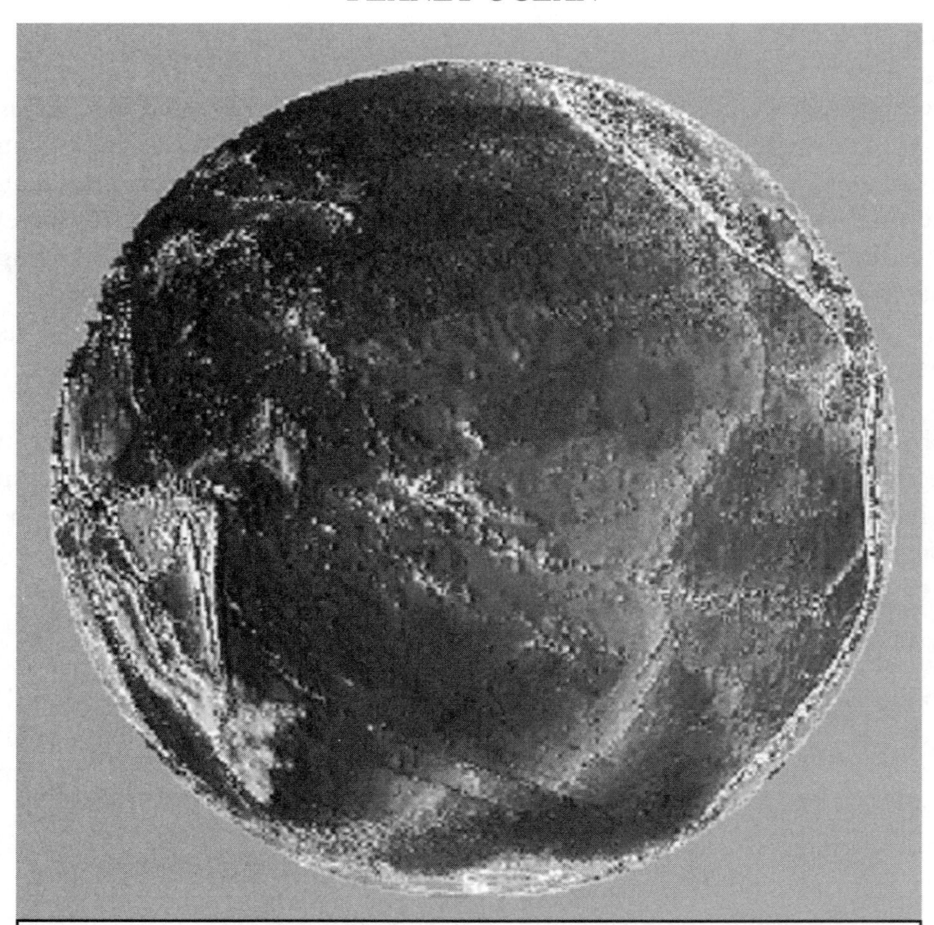

Here we see a view of the Pacific Ocean from 35785 km above 15° S 140° W. This global topographic map is courtesy of the Marine Geology and Geophysics Division of the National Geophysical Data Center operated by the United States Department of Commerce, National Oceanic and Atmospheric Administration.

The Pacific Ocean is the largest single feature on Earth. Its area is comprised of about 66 million square miles. Looking at the Pacific Ocean from space, we see that it makes up almost

1/2 OF THIS PLANET!

Genesis 1:1-2
"In the beginning God created the heaven and the earth.
"And the earth was without form, and void;
and darkness was upon the face of the
deep. And the spirit of God moved
upon the face of the waters."

The Pacific Ocean encompasses an area larger than all the land surfaces on earth put together. All plants, animals and humans require water to survive. We know that if it weren't for water, nothing on this planet could live. It is no wonder that throughout the Bible, God equates himself with the water. About 97% of the world's water is in the oceans, and the Pacific Ocean is the biggest ocean of all.

Since the Pacific Ocean is so deep and vast, the real contour or shape of its mountain ranges has always been a mystery. But that has now changed with startling results. Modern echo-sounding techniques have led to the most accurate ocean floor maps ever. We will review these techniques and their history, but first let's look at the magnitude of the Pacific and at what must be the most fascinating map of all maps.

Water is the most common yet precious material known. Without it our bodies would wither and die. Water is, without a doubt, the very lifeblood of our world. The Earth has more water than any other planet in our solar system. Scientist Jean Francheteau spoke appropriately when he said, "From the point of view of the earth scientist, our planet probably should be called 'Planet Ocean' rather than 'Planet Earth'."

The Pacific Ocean is the largest ocean on this planet. The total area is about 66 million square miles. Its depth is five times the average height of the mountains contained within it. From high in space we can see that the Pacific Ocean covers almost one-half of our world. In weight, size and volume, the Pacific Ocean, with its unmistakable topography, is the most immense thing on earth.[1]

If we could form all the oceans of the world into one sphere, it would be almost one third as large as the moon. If the world were to become totally flat and the oceans distributed themselves evenly over the earth's surface, the water would be about two miles deep at every point.[2]

THE FACE ON THE PACIFIC OCEAN FLOOR

The dominating feature of this book is this massive profile formed by coastal contours and the topography of the Pacific Ocean Floor. The greatest detail of this profile is seen along the western coastal areas of North and South America. The profile appears to be that of a man weeping and looking at something above. The awesome questions that are raised by this phenomenal reality is what we are attempting to answer.

The Pacific Ocean Floor

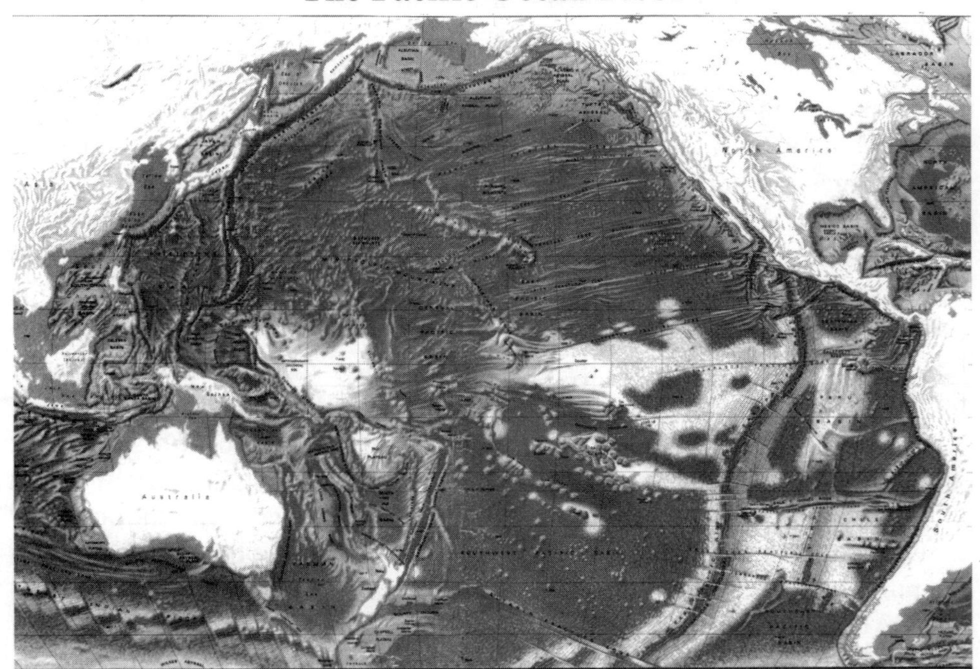

This topographic map of the Pacific Ocean Floor is published by the Rand McNally Map Company.© This map and sections of it are used more than any other map in this book because it is among the most accurate ocean floor maps ever made.

Tracing only where the water meets the land and the largest deep-sea mountain ranges, we highlight the image that is the focus of this review. Visible is the entire right side of a man's head. He appears to be weeping as he looks heavenward with a deep and steady gaze. Exactly who does this personality represent? What could he possibly be looking at? Why are there tears flowing down his face and why is he weeping?

With a typical map of the Pacific Ocean, you cannot see the underwater mountains. To do this, you need a topographic map of the Pacific Ocean Floor. As if from a satellite in space, it's like looking at the entire Pacific Ocean area without the water. When examining this map, first time observers notice something amazing. The face on the Pacific Ocean Floor is even more realistic without a border tracing. That is why the comparison below is with a reduced inset.

The Face on the Pacific Ocean Floor

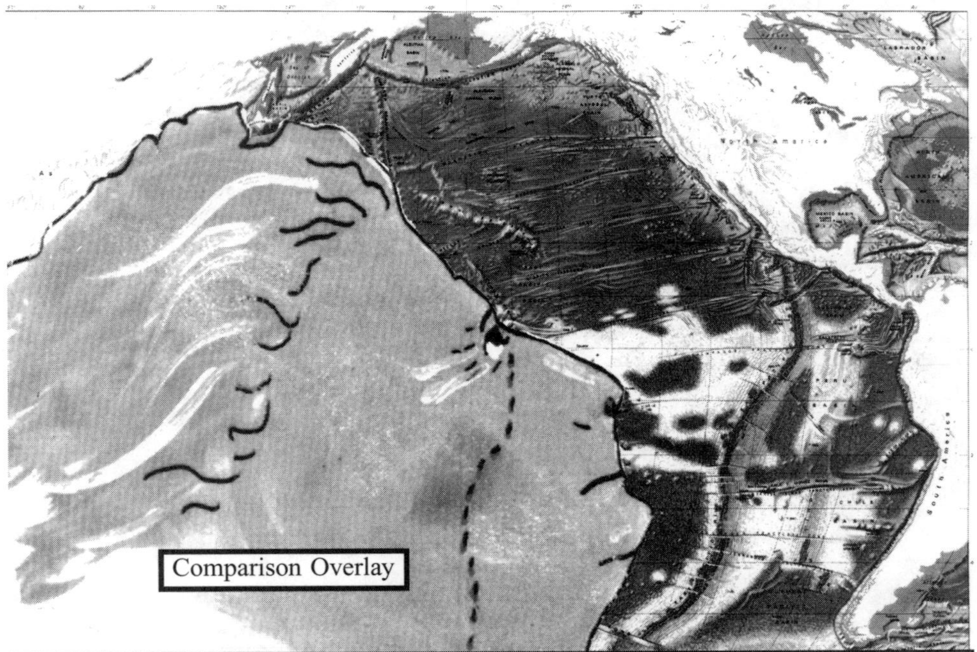

This illustration shows a side by side comparison overlay, as an inset, with the same topographic map that is on the previous page. A more detailed border tracing and comparison of the "face" region can also be seen on the next four pages.

Let us now examine at close range, the most significant areas of this Rand McNally map.[3] We will use various overlays and other maps to further study the details of the mystery of this face on the Pacific Ocean Floor. Our next step will be to examine the right side of this ocean floor map, which highlights the greatest details of the man's weeping face. The comparisons of these first few pages are a helpful foundation for the rest of this book.

With an Onionskin Overlay

Here a felt tipped pen was used, with onion skin paper, to trace along the continental borders of North and South America on the Rand McNally map of the Pacific Ocean Floor.© Also traced is the eye and tear-track of the "face" region of the weeping man.

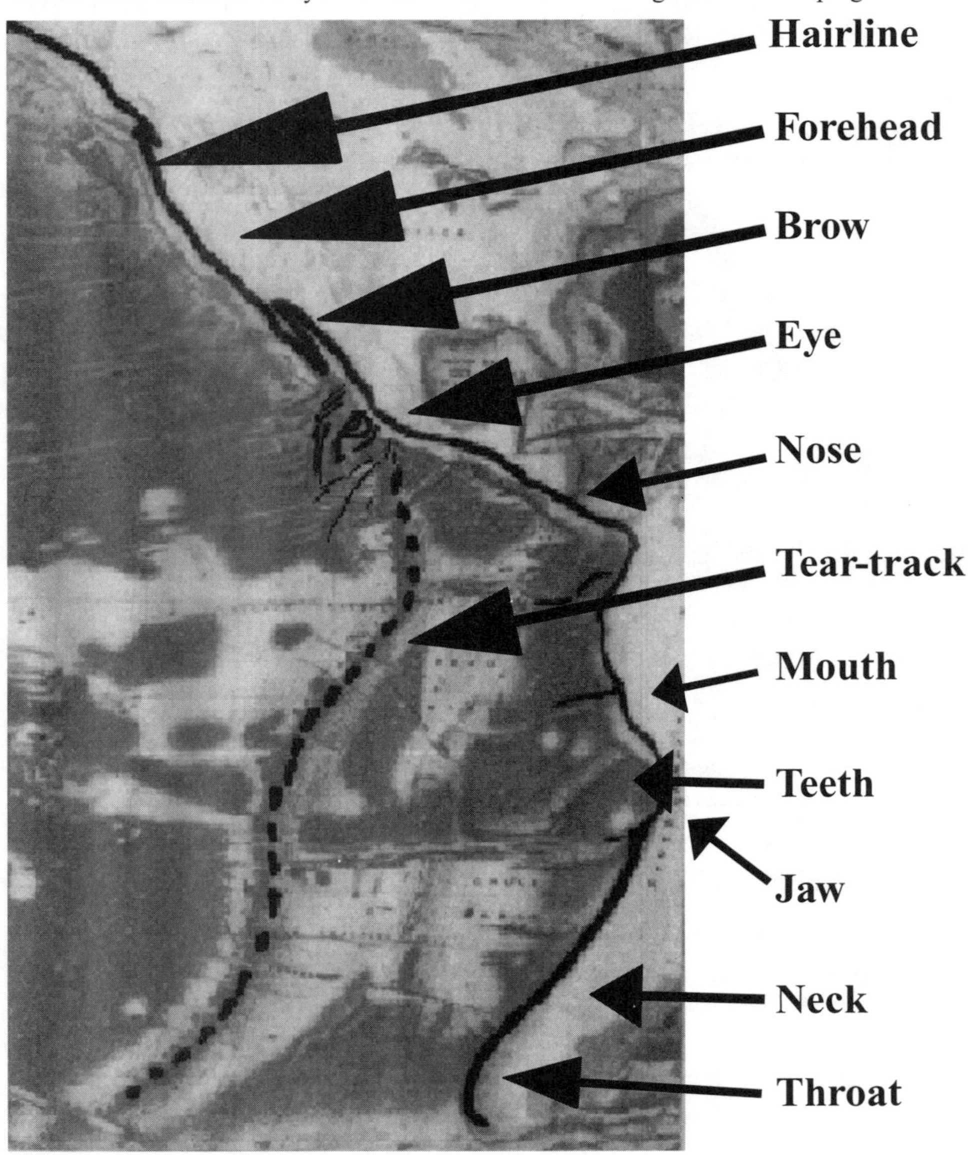

Without an Onionskin Overlay

Here we see the same area as shown on the previous page, without a tracing, of the Rand McNally map of the Pacific Ocean Floor.© Notice that the details of the "face" region of the weeping man are even more realistic appearing on this map without a border tracing.

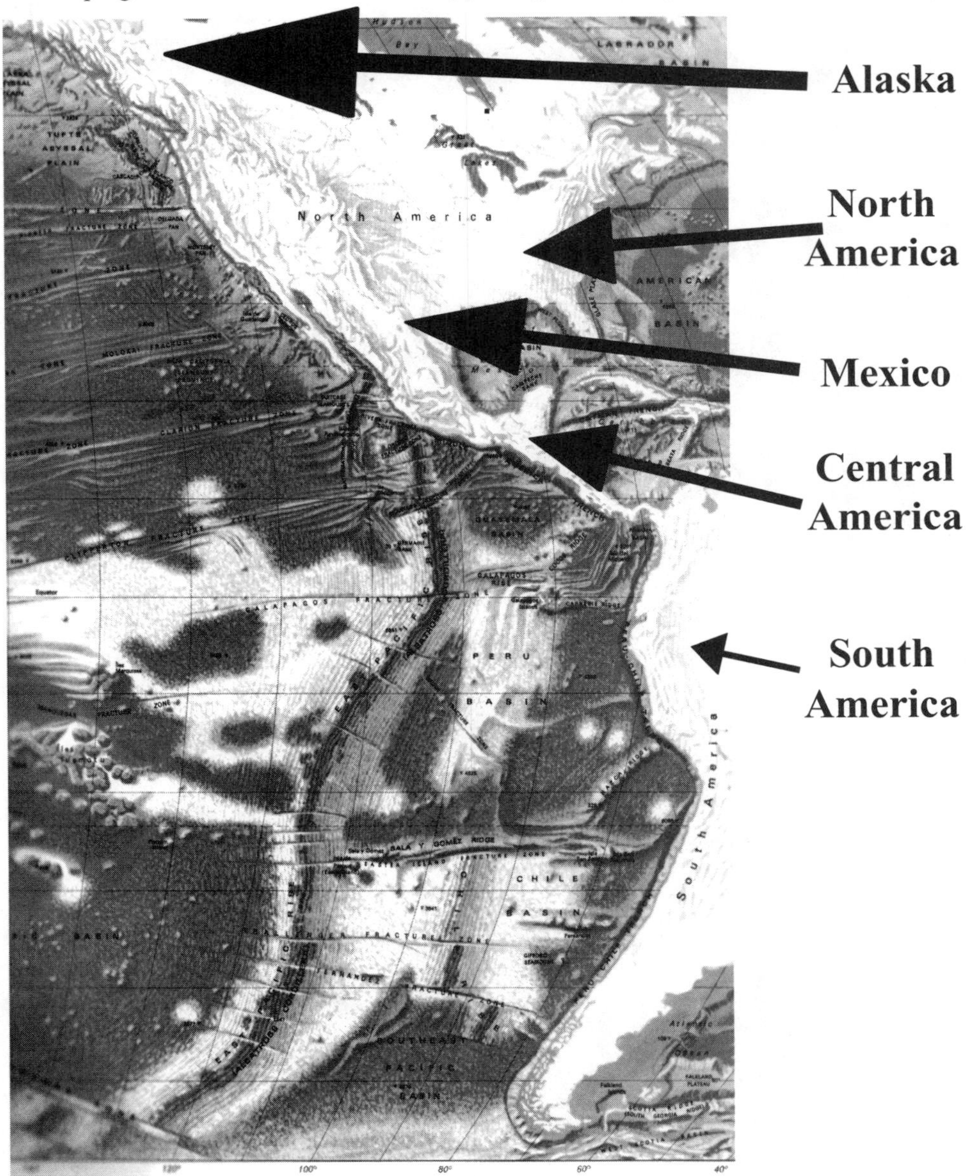

With an Onionskin Overlay

A repeat in greater detail of the tracing on page 10.

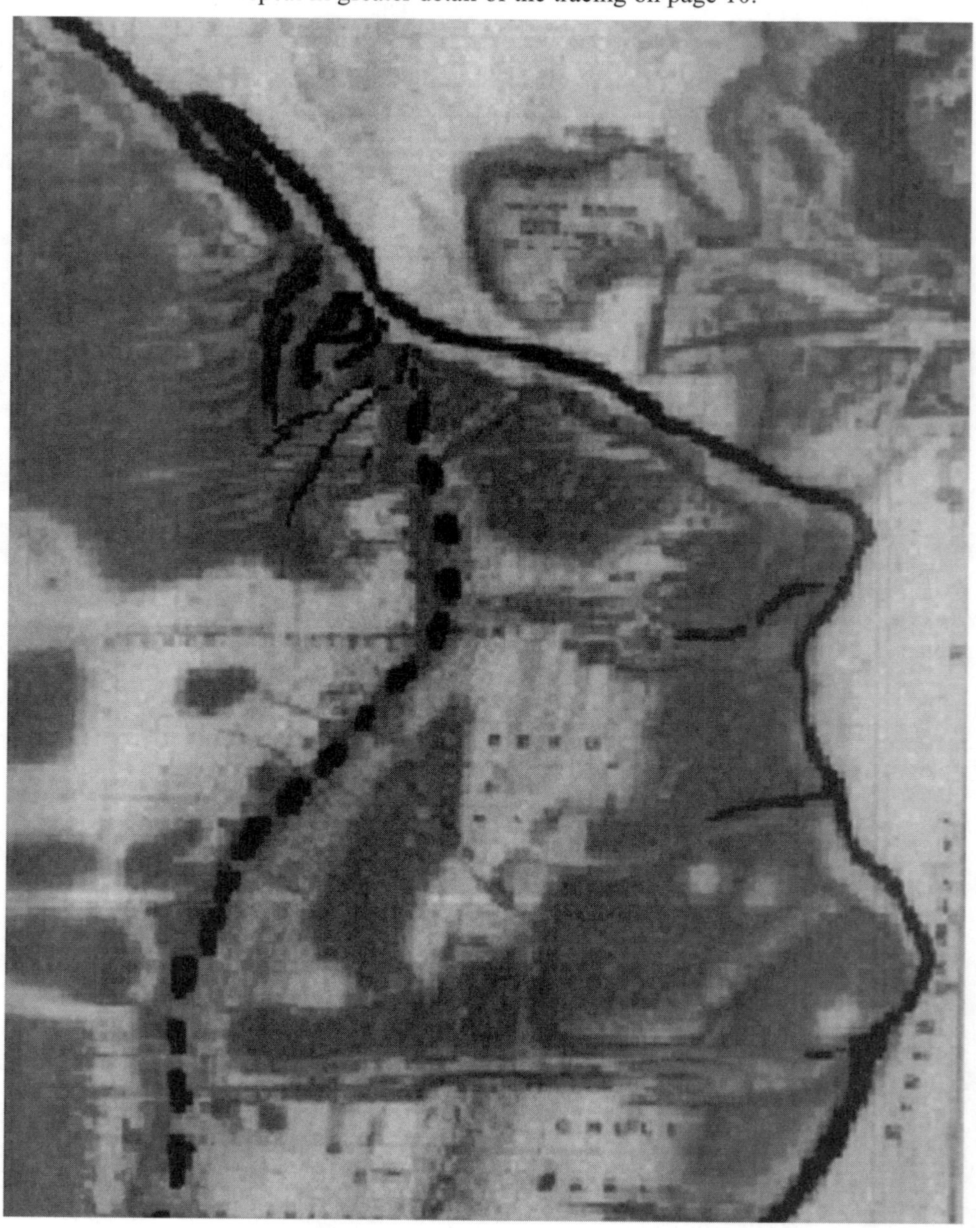

Without an Onionskin Overlay

A repeat in greater detail of the map on page 11.

Comparing the various maps and tracings in this book can be helpful. Before going further, you may want to stop and examine carefully the easily recognizable elements that make up the dominant features of the face on the Pacific Ocean Floor. We are using this map of the Pacific Ocean Floor throughout this book, in full and in sections because we believe it is among the most accurate. If you would like to study a large color enlargement of any of the Rand McNally maps used in this study, an excellent source is "Goode's World Atlas," which is published by Rand McNally & Company.[4]

Many of these maps can also be found in the last few editions of "The Random House Encyclopedia."[5] Both of these volumes are available at bookstores almost everywhere. Hundreds and maybe even thousands of people were involved in the work required to compile this map. Rand McNally explains that the designing of it, as with all their ocean floor maps, was well thought out. This rendering attempts to "convey an impression of the actual physical nature of the ocean floor.[6]

COLOR AND SHADING MATTER

In general, the colors in use with these maps are similar to those on the actual ocean floor. For continental shelves or shallow inland seas, the grayish-green corresponds to terrigenous oozes and sediments washed from the continental areas. In deeper parts of the oceans, calcarous oozes (derived from the skeletons of marine life) appear in white. The fine mud from land is red. In the Pacific, slower sedimentation results in an abundance of manganese and hence darker colors.[7]

Undersea ridges in black suggest recent upwelling of molten rock. Small salt and pepper patches portray areas where manganese nodules are found. Around many islands, white is to show coral reefs. We see differences in relief with the use of relief shading.[8] I like to put it this way: If it were possible to drain all of the water from the Pacific Ocean and then examine it from a satellite in space, it would look very much like this map. That's how accurate it is. The colors might be a little different, but the various shapes would be the same.

Some people might be curious about how these topographic maps are made. If these really are the most accurate ocean floor maps ever made, how do scientists get the data required to do their work?

Scientists measure the ocean's depth by timing the echo of sound waves bounced off the ocean floor from ships. For the first time in recorded history we can see a true picture of the great ocean basins. Oceanographic scientists gather their mathematical data from the measurements of specially calibrated sound waves. These sound waves can go out several hundred miles. Then they measure the final echo results. They transcribe this information onto graph paper and use it to design modern ocean floor maps.

UNKNOWN DEPTHS

The depth of the ocean was a complete mystery to ancient mariners. Many scientists thought that the ocean bottom was flat and level. In 1521, Magellan attempted the first deep sea soundings. He tied a cannonball to a long rope in an attempt to measure the ocean floor. His line was only a little more than five hundred feet in length, so it never even touched bottom.[9]

Then came the deep-sea soundings of Sir James Ross and M.F. Maury in 1840. By 1854, they were producing the first bathymetric charts. These early soundings were not done with sonar. Scientists would undertake the difficult process of using wire or rope and each sounding took several hours. As late as 1912, there were only about 5,969 soundings in depths of 1,800 meters (100 fathoms) or greater.[10]

There are several very deep areas in the Pacific. Some of these include the Tonga and Kermadec trenches to the west of the South Fiji Basin. Both are more than 10,000 meters. Similar depths can be found in the Japan Trench and Kuril Trench, near the Japanese Islands. Scientists know that there are more than fifty known ocean depths greater than 6,000 meters.[11]

In 1952, oceanographer H.M.S. Cook recorded the deepest electronic sounding ever recorded. "This took place in the Mindanao Trench, east of the Philippine Islands. The echo-sounder registered 11,515 meters or over 7 miles. If Mount Everest (29,141 feet), the world's highest mountain, were dropped into the Mindanao Trench, it would be covered by water to a depth of about 1.6 miles."[12] For more in-depth scientific information relating to the maps we use, we suggest that the reader refer to the atlases and the related books that we footnote.

A CLOSER LOOK AT OUR CRITERIA

Where the water meets the continents

Exactly what is our criteria concerning the tracing of major images on any ocean floor, such as the face on the Pacific Ocean Floor? *This study is primarily interested in images seen through ocean border tracings, traced where the water meets the continents.*

Also, we use general tracings of the highest mountains and the deepest crevices on the ocean's floor. How limiting is this criteria for recognizing images from topographic maps? Notice that the tracing lines are over the edges of major land mass continents. By using this as our criteria, we severely restrict how we can interpret what we see. Even a child tracing this map along its continental borders will arrive at images similar to ours.

For example, when tracing the face on the Pacific Ocean, the first consideration is to trace the line beginning at the tip of South America. Then we trace along its western border. We continue tracing the western borders of Central America, Mexico and North America without breaking a line. This line continues and eventually becomes the Leutian Trench, Kamchatka Trench, Kuril Trench, and Japan Trench. Then the line splits. It surrounds the Philippine Basin and extends on through the South China Sea.

Our goal is not to interpret or guess where the tracing lines should begin or end. Tracing guidelines are dictated by specifics. We trace only the *outer edges* and *highest underwater topography* of the Pacific Ocean Floor.

This simple tracing makes up the image of a man's head. He has a giant tear-track going down the right side of his face. In actuality, this tear-track is the East Pacific Rise. Our tracing is free-style and the intention is not to produce a carbon copy. For example, our tracing of the eye is more to guide the reader in placement than anything else. The actual eye area on the map has much more detail than our tracing. On the Rand McNally map, we can see clear separation of all the important parts of the eye, from the pupil to the sclera. This even includes highlights and shading.

With continued study of this map we find some interesting comparisons. We are able to see the exterior right side view of a man's head, but *also comparative features inside,* like the "Pineal Gland." It lies in the middle of the Pacific at the Line Islands near Christmas Ridge.

Also visible are facial bones, as well as teeth located inside the mouth at the Nazca Ridge. It requires little imagination to see some of his right ear, near the Fiji Plateau. But more than observing any physical presentation is the spiritual feeling one can sense when pondering the complexity of this image on the Pacific Ocean Floor.

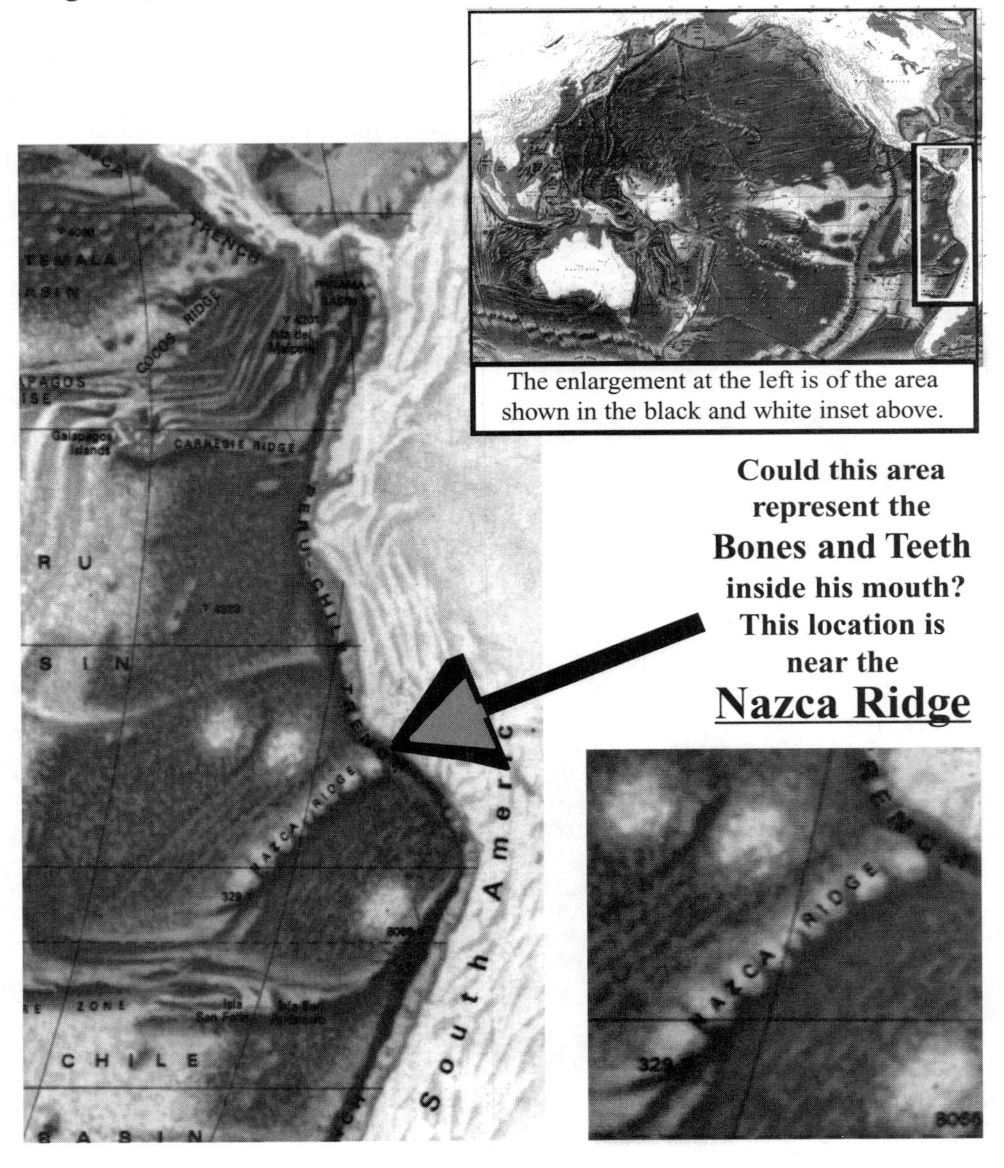

The enlargement at the left is of the area shown in the black and white inset above.

Could this area represent the

Bones and Teeth

inside his mouth? This location is near the

<u>Nazca Ridge</u>

THE TEAR-TRACK

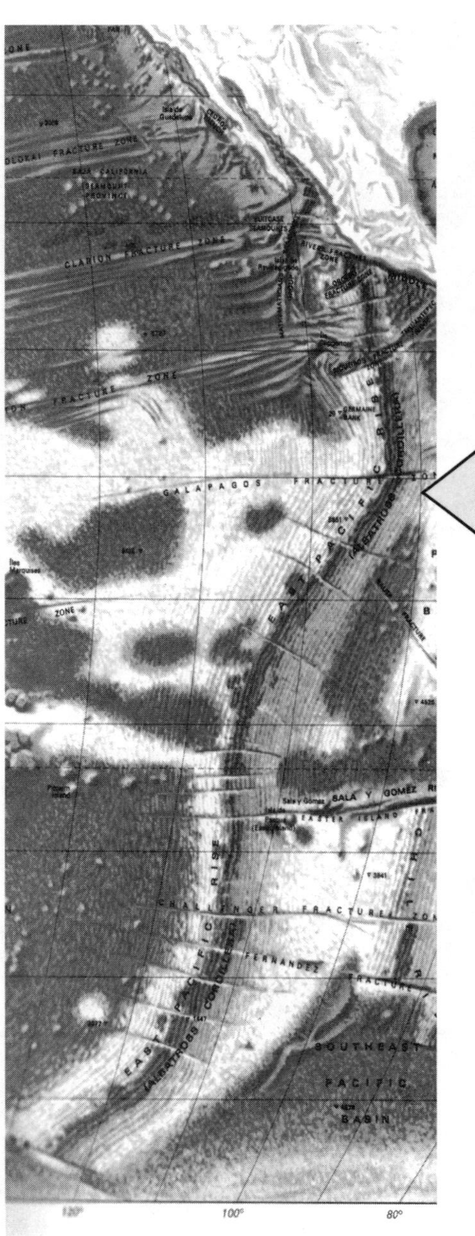

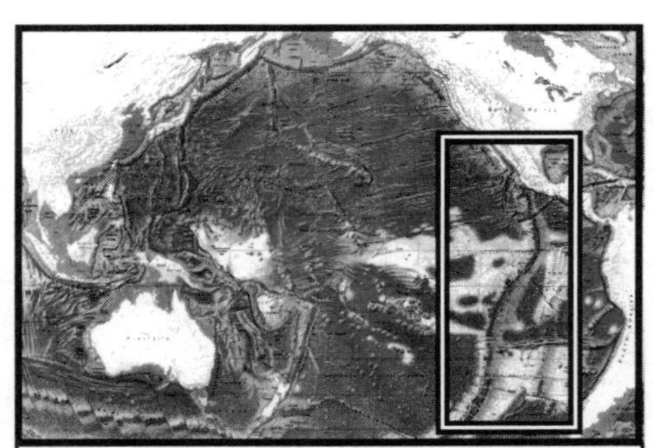

The enlargement at the left is of the area above in the black and white square.

The East Pacific Rise

Highlighting the <u>TEAR-TRACK</u>

The
East Pacific Rise
is part of the
Largest
Mountain Range
on
Earth.

Chapter 2

THE EYE OF GOD?

THE TEAR-TRACK

Face section of the Rand McNally Map of the Pacific Ocean Floor.©

The East Pacific Rise

The most predominant of all the underwater mountain ranges on the Pacific Ocean Floor is the gigantic East Pacific Rise. We call this a "tear-track" on the face profile. It begins adjacent to Manzanillo, Mexico, and then continues to meander for several thousand miles south to the huge Eltanin Fracture Zone.

The tear-track begins at the tear duct of the eye. This starts at the "Rivera Fracture Zone" which is also just off the coast near Manzanillo, Mexico. The East Pacific Rise is where the fastest rates of plate-spreading are found. It runs along north into the Gulf of California, disappearing under the western United States where it connects on land with the San Andreas fault.

What Is He Looking At?

This is the face section of the Rand McNally Map of the Pacific Ocean Floor.©
The above square shows the "Eye Area" of the enlargement on the next page.

THE EYE AREA

Looking at a topographic map of the Pacific Ocean Floor is an event. It can be an astounding experience to some and a shock to others. Much of this comes from the unrelenting evidence off the west coast of Mexico. His weeping right eye is unmistakable. This is the area of greatest clarity and is where all first-time skeptics should direct their attention.When compared with the face section of the map on the previous page, it is easier to recognize that this certainly does appear to be an eye.

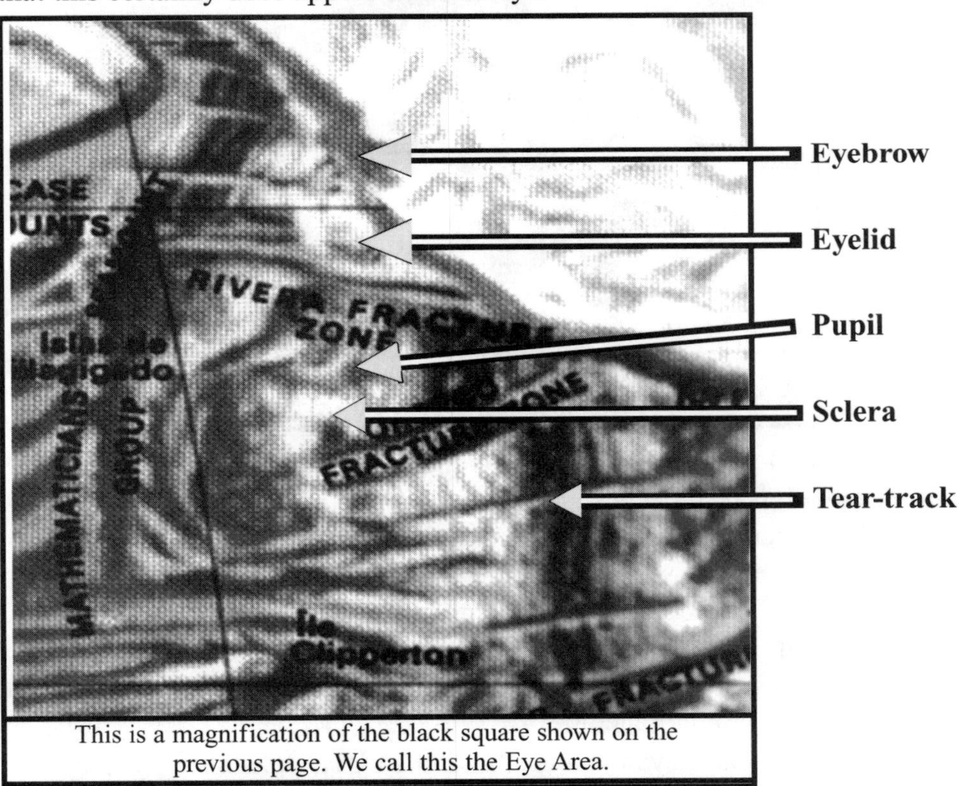

This is a magnification of the black square shown on the previous page. We call this the Eye Area.

The image shows only his right eye which is located off the coast of Manzanillo, Mexico. It is colossal. Its area is greater in size than the cities of Los Angeles, New York and Chicago combined. Since there has been speculation that this face might represent the God of the Bible, then could this eye represent the eye of God? The subject of the eye of God must be extremely important. Let's look at this in various cultures.

TRADITIONS ABOUT THE EYE OF GOD

The idea of the "Eye of God" is also important in religions other than Christianity. It is a subject that has universal religious significance throughout antiquity. In ancient Egypt, the Sun was considered the right eye of their all powerful God, named "Ra." They believed that in the right eye of God were united the brilliance, power and very life of humanity.[1]

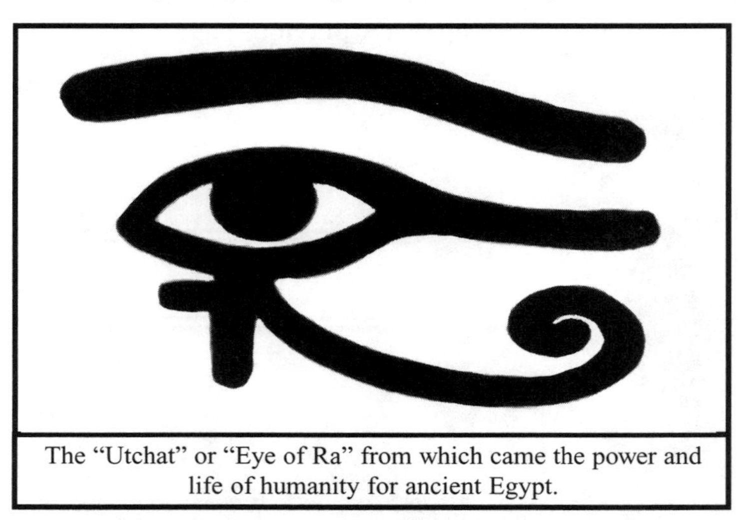

The "Utchat" or "Eye of Ra" from which came the power and life of humanity for ancient Egypt.

In the Egyptian legend of Shu and Tefnut, the two are watched over and eventually saved because of the Eye of God. The text reads, "My Eye was behind them during the aeons that they passed far away from me." A variant would be, "It was my Eye that brought them back to me after the aeons that they stayed far away from me." The "Utchat" or "Eye of Ra" and sometimes called the "Eye of Horus," the hawk, was considered all powerful to the Egyptians. In some stories we see the eye blinded or growing weak and then gaining its strength again.[2]

The ancient Mayan text the "POPOL VUH," tells of two young men. They are entering a court to play ball. Upon arriving a hawk appears on the scene and begins cawing, "Vacco! Vacco!" meaning, "Here is the hawk! Here is the hawk!" Out of fear, one young man shoots a pellet at the eye of the hawk and the bird falls to the earth blinded. When the hawk promises a message, they anoint his eyes with rubber. The hawk's eye recovers and healing is instantaneous.[3]

"The eyes are next in nature unto the soul," wrote the philosopher Dr. Culpeper, "for in the eyes are, seen and known, the disturbances and griefs and gladness and joys of the soul." He adds, "They are to the visage that which the visage is to the body; they are the face of the face; and because they are tender, delicate and precious, they are fenced on all sides with skins, lids, brows and hair."[4]

Author Manly Hall writes that "Leonardo da Vinci believed there were certain quaint virtues emanating from the eye. He (Leonardo) explains that a snake called the 'Lamia' has a magnetism in its gaze which draws the nightingale to her death." He also notes that "The wolf has the power to turn men's voices hoarse with a single look; that the basilisk can destroy every living thing with its glance; that the ostrich and spider hatch their eggs by looking at them; and that a fish called 'linno' (found off the coast of Sardinia) sheds light from its eyes. When it does, all fishes which come within its radiance, die."[5]

One thing intrigues many students. They notice that when observing the face on the Pacific Ocean Floor, what's actually visible is the entire right side of the head. The clearest part, the part that shows the *most identifiable detail and recognizable personality, is the right eye.*

BIBLICAL REFERENCES TO THE "EYES"

In the Bible, there are references to God's eyes. There are also references to his "right" eye. Often the "single" eye, is synonymous with Godliness and perfection. But, we *do not see the left eye referred to specifically as an identifying attribute relating to God.* Also, the scriptures show that the eye not only absorbs light but it also emits light. We see evidence of this in:

Matthew 6:22
"The light of the body is the eye: if therefore thine eye be single, thy whole body shall be full of light."

David had aspirations. His wish was to be the "apple" of God's eye.[6] When Ezekiel sees the vengeance of God against Israel, he describes the power of the Lord as coming from God's eye. After promising vengeance,

God says “and mine eye shall not spare thee.”[7] Matthew quotes Jesus as advising that an offensive eye is such a hindrance, that it should be plucked out and thrown away.

Matthew 18:9
“And if thine eye offend thee, pluck it out,
and cast it from thee: it is better for
thee to enter into life with one eye,
rather than having two eyes
to be cast into hell fire.”

The prophet Paul wrote that when the world ends, Christians will not die. They will change instantly and dramatically. Those of us that are saved from condemnation will experience a new power and strength. We will become immortal and incorruptible. Paul writes that this will happen “in the twinkling of an eye.”[8] The Bible also shows that human vision has been partly blinded since the original fall of man. In the “Garden of Eden,” the serpent tries to convince Eve to eat the forbidden fruit. He says that she will not die and her eyes will be opened.

Genesis 3:5
“For God doth know that in the day ye eat
thereof, then your eyes shall be opened,
and ye shall be as gods, knowing
good and evil.”

Before Adam and Eve ate the forbidden fruit, the Bible states that this fruit was “pleasant to the eyes.” Once they ate this fruit, they were *able to see differently*. For the first time in their lives, they both knew shame.

Genesis 3:7
“And the eyes of them both were opened and
they knew that they were naked; and they
sewed fig leaves together, and
made themselves aprons.”

LOOKING THROUGH A GLASS DARKLY

Great things happen in scripture surrounding certain individuals and their ability to see or not. *It may sound strange, but Adam and Eve experienced an opening of their eyes. Since then, there has been a prevailing blindness in all humanity.* Saul killed Christians in the name of God. Then God blinded him so he could see that his ways were wrong and be converted (Acts 13:11). Later, Saul, who changed his name to Paul, further substantiates the blindness of all people. In the last days, humanity will return to a perfect kind of wisdom and vision.

I Corinthians 13:12
"For now we see through a glass, darkly; but then face to face: now I know in part; but then shall I know even as also I am known."

Dante also makes reference to the imperfect vision of all mankind in Canto II of his Purgatory. "What negligence detains you loitering here? Run to the mountain to cast off those scales, that from your eyes the sight of God conceal."[9]

Dante also wrote a prayer to Jesus. It begs the Lord to turn his "just eyes" back on a lost human race. "If it be lawful, Oh, Almighty Power! Who was in earth for our sakes crucified, are thy just eyes turn'd elsewhere? Or, is this a preparation, in the wondrous depth of thy sage counsel made, for some good end, entirely from our reach of thought cut off?"[10]

Chapter 3

WEEPING AND TEARS

John 11:35
"Jesus wept."

The face on the Pacific Ocean Floor is weeping. Does this face represent a recognizable manifestation of God the Creator? What does the Holy Bible teach about God's weeping? Are Christians weak for loving a weeping God? Does the fact that God weeps, reveal a sign of weakness on his part? What does this part of God's nature tell us about ourselves? It is the love and mercy from God that results in his weeping. It is from his mercy and our faith that we are saved (Matt. 9:22).

When God made human beings, he decided that they should have the same basic features and characteristics as himself. This lies behind the expression that mankind is created in the "image of God" (Gen. 1:26). Even though God is a holy spirit (Jn. 4:24), sometimes he represents himself to his creation with human-like features (Gen. 1:27). The Bible speaks of God's heart, hands, feet and eyes. God weeps and humans also weep. Even if such examples are simply metaphorical, we cannot escape the fact that such anthropomorphic metaphors are common in the Bible.

Genesis 1:27
"So God created man in his own image,
in the image of God created he him;
male and female created he them."

The first thing a baby does when he is born is cry. Weeping is a human trait. Dogs, cats, horses, cows, mice and sheep do not shed tears the way we do. Even apes don't weep in the way that humans weep. Humans weep at funerals, weddings and other events. We live on a planet with humans that uniquely weep. That same planet reveals a massive face on the Pacific Ocean Floor that also "weeps."

God is not unlike the "Teacher of Righteousness" quoted in the Dead Sea Scrolls. The scroll reads, "My eyes are like fire in the furnace and my tears like rivers of water."[1] Throughout history, many images and icons are said to weep. Numerous statues of Jesus have been reported to weep. Throughout the world, Catholics carry out pilgrimages to view paintings and statues showing the Virgin Mary weeping. Ancient Masonry depicted a weeping virgin as the symbol of Virgo.[2] In the "Epic of Gilgamesh," Ishtar Queen of Heaven reminisces, "The great gods of heaven and hell wept, they covered their mouths."[3]

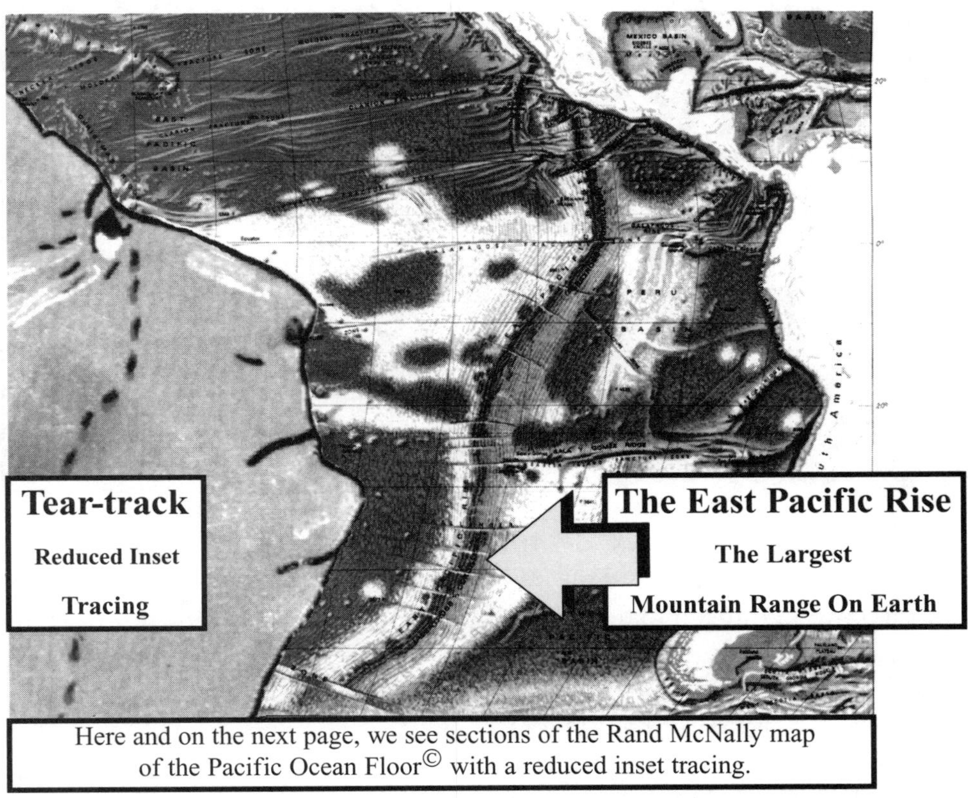

Here and on the next page, we see sections of the Rand McNally map of the Pacific Ocean Floor© with a reduced inset tracing.

Specifically, what are tears? According to the Random House Encyclopedia, tears are: "A fluid that moistens the surface of the cornea of the eye, secreted by the lacrimal glands. It is antibacterial in nature and improves the optical properties of the eye by forming a thin, smooth film on the cornea and compensating for slight surface imperfections."[4]

The Ancient Egyptians taught that human beings were created from the tears of a weeping God. It was said that the Creator wept, and from "his tears (rmyt), humanity (rmtt) was born."[5] Tears were as "bread" in spiritual substance to the Egyptians. This is found in the so called "Egyptian Book of The Dead," more aptly called "The Book of the Great Awakening."[6]

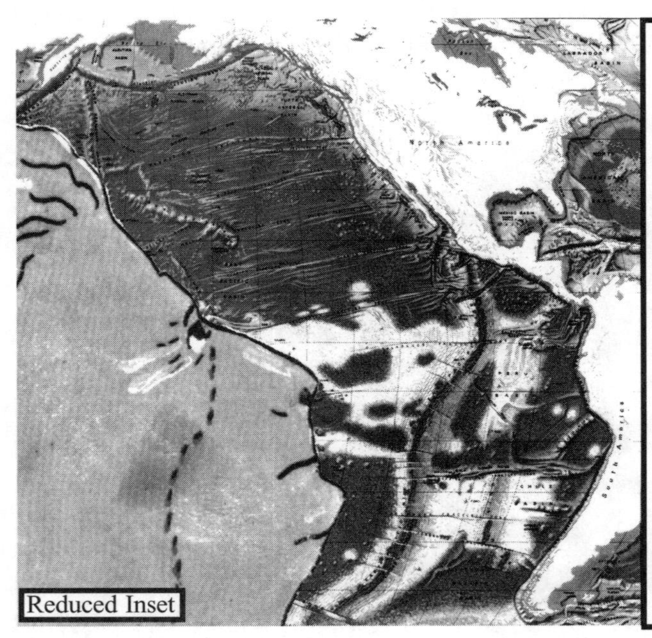

Why Is He Weeping?

Notice all of the many key similarities that there are between our reduced tracing inset and the face area of the topographic map of the Pacific Ocean Floor. If this image does represent God, communicating with all of humanity, then what type of emotions might he convey? His tear-track alone shows attention, warmth, mercy, compassion, strength and watchfulness. As this study will establish, what he is looking at, the object of his affection, has everything to do with why he is weeping.

A section of the text of "Unas" devotes itself to a liturgy involving spiritual resurrection, called the ceremony of "Opening the Mouth." Part of the goal was to transfer to the deceased, among other things, the power of the Eye of Horus.[7] The ancient text reads, "...and he comes and brings to the gods the bread that he finds there, that is, the tears of the Eye of Horus on the foliage of the Tchenu tree." The "Eye of Horus" as introduced in the last chapter, was as renowned for its weeping as well as its shape. It is one of the most well known hieroglyphs from ancient Egypt. It is commonly seen among the artifacts on display at the Cairo Museum. Always protruding from the sty of the "Eye of Horus" (see page 22) is a tear-track and a teardrop. The "Eye of Horus" served as a reminder to the ancient Egyptians that their god *wept* them into existence.[8]

There is minimal ocean floor detail seen on the computer generated topographic map below. For example, the East Pacific Rise, which we also call the tear-track is less clear. But from this viewpoint, we can see a clear outline of the west coast of North and South America and the

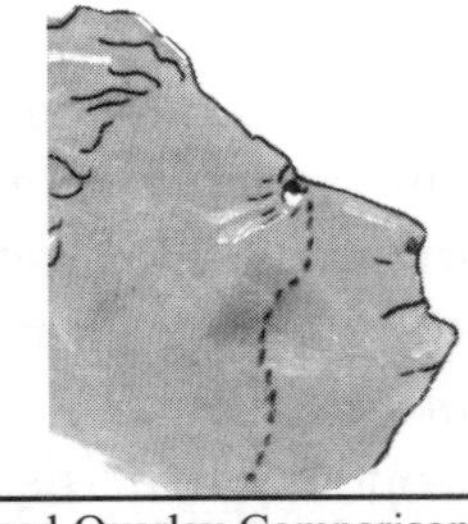

Reduced Overlay Comparison

Pacific Ocean

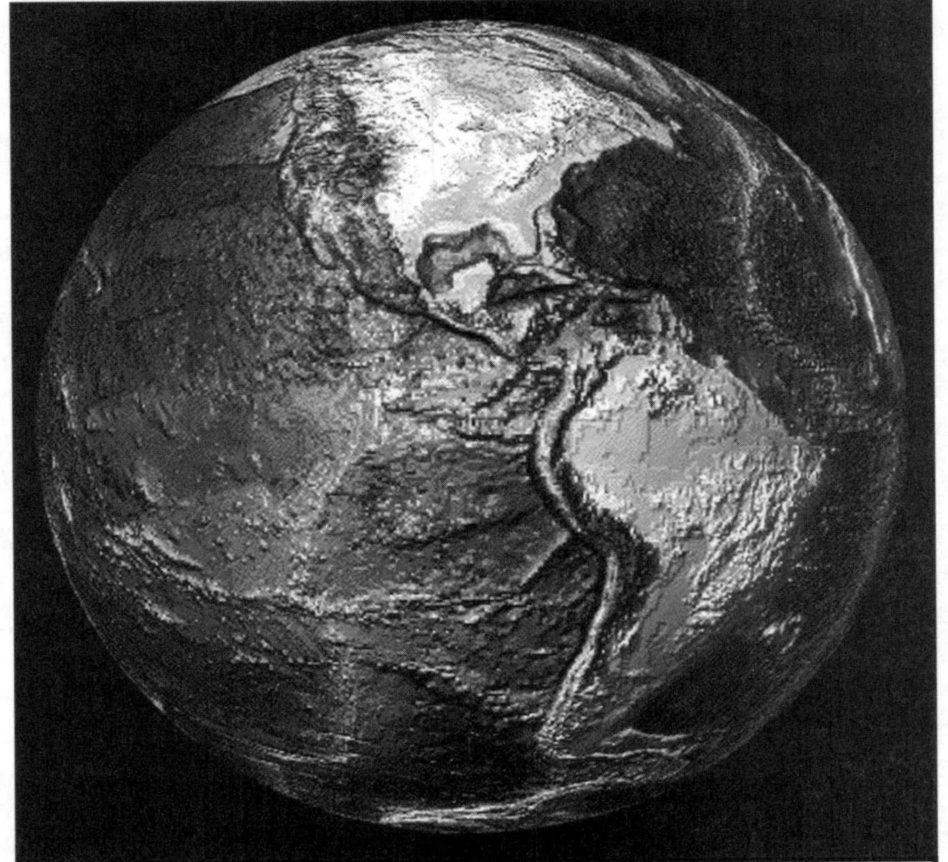

This global view, highlighting the face and tear-track area of the face on the Pacific Ocean Floor, is from the U.S. National Geophysical Data Center. The shading and contrast were chosen to give a natural look to the continents and oceans according to elevation. Major sources include the U.S. Oceanographic Office, the U.S. Defense Mapping Agency, the Bureau of Mineral Resources and NASA.

AN OCEAN FULL OF TEARS

The Egyptians, Greeks and Latins, highly esteemed tears and collected them in lachrimatories, or tear bottles. Sometimes these tear bottles were buried with the dead as evidence of grief. More often, they were preserved for purposes of healing or magic.[9] Among other things, tears make us see more clearly. Tears are (on page 27) defined as a fluid that "also improves the optical properties of the eye by forming a thin, smooth film on the cornea and compensating for slight surface imperfections."

THE GOODNESS OF TEARS

A good cry has always been known to heal the body and spirit. Now we know that by the very result of tears, we also see more clearly. We live on a planet that is literally *filled with an ocean of tears.* We see this emulated by studying the huge weeping face on the Pacific Ocean Floor. It is also a chemical fact. Scientists have long realized that human tears are similar in salinity (salt content) to sea water.[10]

Tears are very salty and so is sea water. There is an estimate that if all the seas of the earth dried up, they would leave about 4,419,330 cubic miles of rock salt. This would be enough salt to cover the entire United States with a layer one and a half miles deep.[11]

WEEPING IN THE BIBLE

The shortest verse in the Bible has only two simple but powerful words. One of them is "*Jesus*" and the other is "*wept*" (Jn. 11:35). In a unique way, it could be said that the only word in the Bible that holds equal status to the word *"Jesus"* is the word *"wept."* The first time we view the word "wept" in the Bible (Gen. 21:16), it results in the young man Ishmael being saved from death. The shortest verse in the Holy Bible, *"Jesus wept,"* results in a similar scenario. The young man Lazarus is raised from the dead.

The first time we see weeping in the Bible is in the book of Genesis (21:16). God found it fitting that this first weeping event recorded would not come from a prophet or even from himself. It would come from a slave who was dying of thirst and worried about the life of her only son. Hagar, the bondwomen who bore Ishmael, the son of Abraham, cries out to God. It was at the absolute crossroads of her life, a time when all seemed hopeless.

After Abraham sends Hagar away with her son, she wanders in the wilderness of Beersheba. Ishmael is so close to death that Hagar sits a good way off from him and grieves. Then, unable to contain herself, she lifted her voice and wept (v16). Then the Angel of God (v17) hears her prayer and the groaning of her son and leads Hagar to water.

The subject of "weeping" or "tears," occurs hundreds of times throughout the scriptures. When Abraham's wife Sarah died in Hebron, in the land of Canaan, the scriptures tell us that *"he came to mourn for Sarah and to weep for her."*[12] When Jacob's brother Esau begged his father Isaac for a blessing, the Bible says that he *"lifted up his voice and wept."*[13]

When Jacob met his future wife Rachel, he kissed her and *"lifted up his voice and wept."*[14] His gratitude to God was overwhelming. Later Jacob fears that his brother Esau will kill him, so he bows himself to the ground seven times. In response to this gesture, Esau runs to meet his brother, embraces him and falls on his neck. Then Esau *"kissed him: and they wept."*[15] This is the first time in scripture that we see two people weeping together.

At the end of Genesis, when the brothers of Joseph begged him in the name of their father to forgive them, Joseph wept.[16] The "Weeping Prophet," Jeremiah said that his soul would weep for Israel and its pride. Then he wrote that his eyes shall "*weep sore and run down with tears."*[17] Many people believe that his lament was because of the captivity of Israel by the enemy nation of Babylon. Jeremiah, in an earlier chapter, shows his desire to emulate the total mercy of God. We see this in:

Jeremiah 9:1
"Oh that my head were waters, and mine eyes a fountain of tears, that I might weep day and night for the slain of the daughter of my people!"

A few verses later Jeremiah states a prophecy about mountains and how there will be a time in the future when he will go to the mountains and weep, *"For the mountains will I take up a weeping and wailing"* (v10). He encourages others to *"take up wailing for us, that our eyes may run down with tears and our eyelids gush out with waters"* (v18). In the book of Psalm, King David describes his tear-drenched bed.

Psalms 6:6
"...all the night make I my bed to swim;
I water my couch with my tears."

Then David justifies his power against the workers of iniquity; *"for the Lord hath heard the voice of my weeping"* (v8). Here the Bible teaches that weeping results with a response from God. In the book of Luke, we can read of the woman who came from behind Jesus to wash his feet with her tears and wipe them with the hairs of her head.[18]

The last two times the word "tears" is mentioned in the Bible is in the book of Revelation. Both verses repeat a great promise that all of humanity can look forward to and pray for. When Jesus Christ comes to Earth and rule the world, it will be a place of living fountains of waters. Also, there will be no more tears, crying or sorrow.

Revelation 7:17
"For the Lamb that is in the midst of the throne shall feed them, and shall lead them unto living fountains of waters: and God shall wipe away all tears from their eyes."

Revelation 21:4
"And God shall wipe away all tears from their eyes; and there shall be no more death, neither sorrow, nor crying, neither shall there be any more pain: for the former things are passed away."

A PRE-INCAN WEEPING GOD

In the country of Bolivia is an ancient site called Tiahuanaco. Here are probably some of the most controversial archaeological ruins on the planet. Even to contemporary archaeologists who laugh at theories of upheaval, Tiahuanaco is a source of mystery, controversy and debate. Even though there is agreement that these megalithic ruins predate the Incas, their exact age and origin is completely unknown.

The Weeping God of Tiahuanaco

At the center of Tiahuanaco is a stone arch, cut from a solid chunk of andesite weighing about twelve tons. It's been cracked by some great force like an earthquake. On the upper portion of this arch is a series of carvings, believed to be a calendar. In the center a figure, holding a staff on each side appears to be weeping. He is known as "the Weeping God."[19]

WEEPING AT THE APOCALYPSE

When many people think of the "Apocalypse" they picture images of a scorched earth (Rev. 8:5-9) with a third of all life on it destroyed. They think of poisoned oceans (Rev. 8:13) and pestilence (Rev. 9:7) and war everywhere. Although these things are predicted during this overall time period, the word "apocalypse" has more to do with its Greek origin (*apokalupsis*) meaning "to reveal that which is hidden."[20] During this time there will also be an era of understanding and knowledge of God's glory concerning these events that God equates with the vastness of the seas. We can see verification of this in:

Habakkuk 2:14
"For the earth shall be filled with the knowledge
of the glory of the Lord, as the
waters cover the sea."

Of course before the great and final end of the world, weeping and tears will have a lot to do with the events that occur. Fear, torment, weeping and wailing (Rev. 18:19) will accompany these cataclysms. Merchants (Rev. 18:11) will weep because of their great loss and so will the unbelievers who will be cast into outer darkness. Jesus Christ said that during this time (Matt. 8:13) there will be "weeping and gnashing of teeth." It will be a time of great pain for humanity.

A FACE THAT WEEPS

The weeping face on the Pacific Ocean Floor has many clues which can help us better understand the mysteries related to why this face is weeping. The giant tear-track streaming down this humongous face tells us something about his feelings. If the discovery of this gargantuan face is to be considered a harbinger to the Apocalypse, then it is easy to see why knowing what the Bible has to say about weeping and tears is so important. After all, the Bible is the most important apocalyptic book there is. Additionally, a cursory glance at weeping gods in other cultures is important. It lets us better realize that the concept of a weeping deity is one that is familiar to many cultures, not just ours. Then maybe we can even better appreciate the ultimate question about this face.

WHY IS THIS FACE WEEPING?

As we will see in future chapters, there are several other images related to other ocean topography. Sooner or later people interested in this discovery begin to wonder about the possible *story* behind the images on our oceans floor. Once we decide that these images make up a story, then we must ask ourselves, what are the details of that story? What high drama is causing the person identified by the face on the Pacific Ocean Floor to gaze so intently? What is the focus of his undistracted stare? Whatever it is, it must be the saddest thing in the world, because it's causing him to weep *oceans of tears.*

Chapter 4

A PICTURE IS WORTH A THOUSAND WORDS

Imagine for a moment that you are God, the Creator who made the earth and everything on it. You, as God the Creator, wants to give a message to everyone on the planet. It is an important message and yet a simple one that will answer many of humanity's greatest age-old questions. Some of these questions are: Is there a God? Is God human-like or only an impersonal force? If God did make us in his own image, what does God look like? If there is a God, what about the Devil? Is the Devil real? What does he look like? How does the Devil feel about God? Why were we born? Is there a Heaven? Is there a Hell? Which of these are the most important things God might want us to know about?

PICTURES COMMUNICATE BETTER THAN WORDS

Questions about eternal matters have perplexed humanity throughout reported history. Untold thousands of books have wrestled with these issues and millions of people have died in wars attempting to solve them. But if you were God and you wanted to resolve these issues for everyone in the world, would you do it with words? Probably not. What about using pictures that could clearly inform humanity of these great truths?

As God, you might decide to offer answers to these great questions through a simple set of pictures that make up a story. The pictures that you use are so simple that scientists and small school children can equally understand them. It makes sense that God would pick a location like the topography of the earth, where they could not be altered. Then the integrity of God's messages would would be maintained.

Where would you, if you were God, put these pictures? Well since you are God, you can put them any where you want, and if you choose, you can make them large enough for the whole world to see. Why not etch these pictures on the surface of the planet? Why not make the images so huge that it is impossible to ignore them or pretend they don't exist? That is exactly the hypothesis of this book.

That God would want to press an imprint of his face on the topography of the Earth is thought provoking, but until more is known, it is best to think of the face on the Pacific Ocean Floor as a metaphor.

WHY NOW?

Our ancestors lacked the modern technology of mapping. Why would God use topographic pictures that can only be revealed in this day and age? Is it because God wants to communicate with modern humanity for a special apocalyptic reason? That is a possibility because the pictures involved can only be understood with the scientific knowledge of today.

How curious are we? Is seeing really the same as believing? Seeing comes before words. A child sees and recognizes before learning to speak. A simple set of images can tell a complex and detailed story that might otherwise require thousands of printed words to tell. A few good pictures can tell a story understood by every culture without making one translation.

Language, by its very nature, is held in suspicion. Although people can be fooled by pictures, the manipulation of words is the most worrisome. Seeking proof, a wise man does not say "tell me," he says "show me." It is easy to see why God might use pictures rather than words. Pictures are harder to manipulate. Scientists and preschool children, of any language, can equally identify a dog as a dog. People recognize dogs, cats, trees, water and mountains on sight. Familiar things like these and even a human body or face are recognizable without words.

PRIVILEGED COMMUNICATION?

Human history is rich in the tradition of privileged communication through pictures. The greatest paintings are those that require hours of study to digest. Often, what is discovered is much more than expected. Sometimes it is even secret or banned religious dogma. Hidden religious messages in paintings were at times thought to be the last resort for keeping the truth alive. Renaissance painters commonly planted hidden pictures within pictures as a reward to those who would spend enough time gazing at their work to find them. There were times in our recent history that it was an exception if painters didn't have some kind of hidden images in their artistic productions.

HIDDEN IMAGES IN PAINTINGS

In the 18th century, many portrait painters seemed obsessed with hiding images of the crucified or resurrected Christ in their works. One can see the bandaged head of Christ (hidden in a white dress) of Sir Joshua Reynolds' famous work "Diana Viscountess Crosbie."[1]

Some viewers can see Christ's stretched torso (drawn within the green drapes) in the background of Gainsborough's famous portrait of Karl Friedrich Abel.[2] More easily visible is an image hidden in the beautiful painting, "Pinkie" by Sir Thomas Lawrence. Imbedded in her pink gown is a large, fully wrapped, sleeping mummy.[3]

Salvador Dali planted many hidden images in his paintings. Some of these included a portrait of Mae West, a dog, a fish, and even a bust of Voltaire. Al Hirschfeld, who draws caricatures of Broadway's stars for the Sunday New York Times continually hides the name "Nina" (his daughter's name) in almost every drawing he makes.[4]

Hidden pictures in art are deliberate. The artist hides images within his creation to reward those that contemplate his work carefully. But *what about a picture that is half the size of our planet,* as in the case of the face on the Pacific Ocean Floor? Does God, like a painter, reward those who contemplate his handiwork? Can our reality evolve to a higher spiritual level from what we learn because of this research? Although an ocean floor map, based on the earth's actual topography, is not the same as a painting, some of its effects are the same. The more an individual studies a topographic map of the Pacific Ocean Floor, the more the links are discovered and a pattern of harmony is revealed.

AMBIGUOUS ART

Then, there is art that is deliberately ambiguous. One example of this type of optical illusion is of a face-vase. Looking at it, one can see the vase, but its edges also looks like the profiles of two faces.

Most people have seen various versions of this famous drawing. Is it a rabbit or a duck? It can be either, depending on how you look at it, but it's difficult to see both images at the same time.

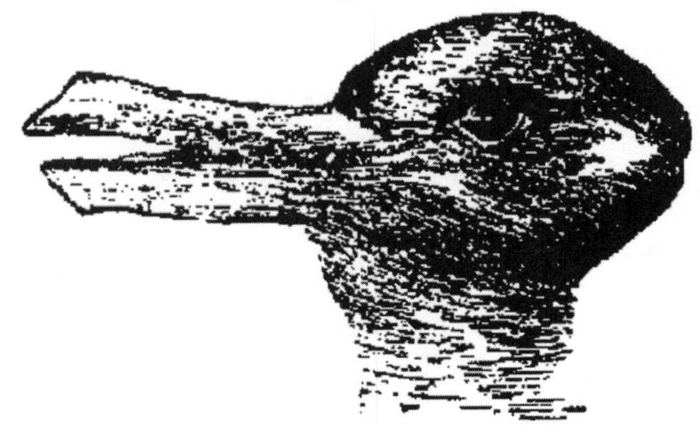

ALL IS NOT AS IT SEEMS IN FRANCE

In 1979, I had the opportunity to visit the Cannes Film Festival. My company was selling a film that I was helping to distribute. Little did I know at the time that during this trip I would learn about a clever optical illusion. One day I found myself attempting to talk with the innkeeper where we lodged. His name was Louis. I spoke no French and he spoke no English, so his wife translated.

She explained that her husband wanted me to print my name on the back of a matchbook. Louis then took out his ink pen and a small ruler. He began busily drawing lines and making little boxes on a plain sheet of paper.

After about three minutes of this, I started to interrupt him because I had to leave. His wife quickly gave me a "shhhh" and smiled like something special was about to happen. She was right. On the next page is an enlargement of the squares and lines that Louis made for me that day. On the surface, they seem meaningless. If you hold this page out in front of you (like in this drawing), the mystery solves itself.

To examine the illusion on the next page, hold it like this.

Hold this page the same way as shown on the previous page, slightly at and angle. When we do, we see that Louis has written my first and last name, “LLOYD CARPENTER” all in capital letters. This seems amazing enough, but when we rotate the page, (one-forth circle to the left) we see that it reads “FESTIVAL DE CANNE(S).”

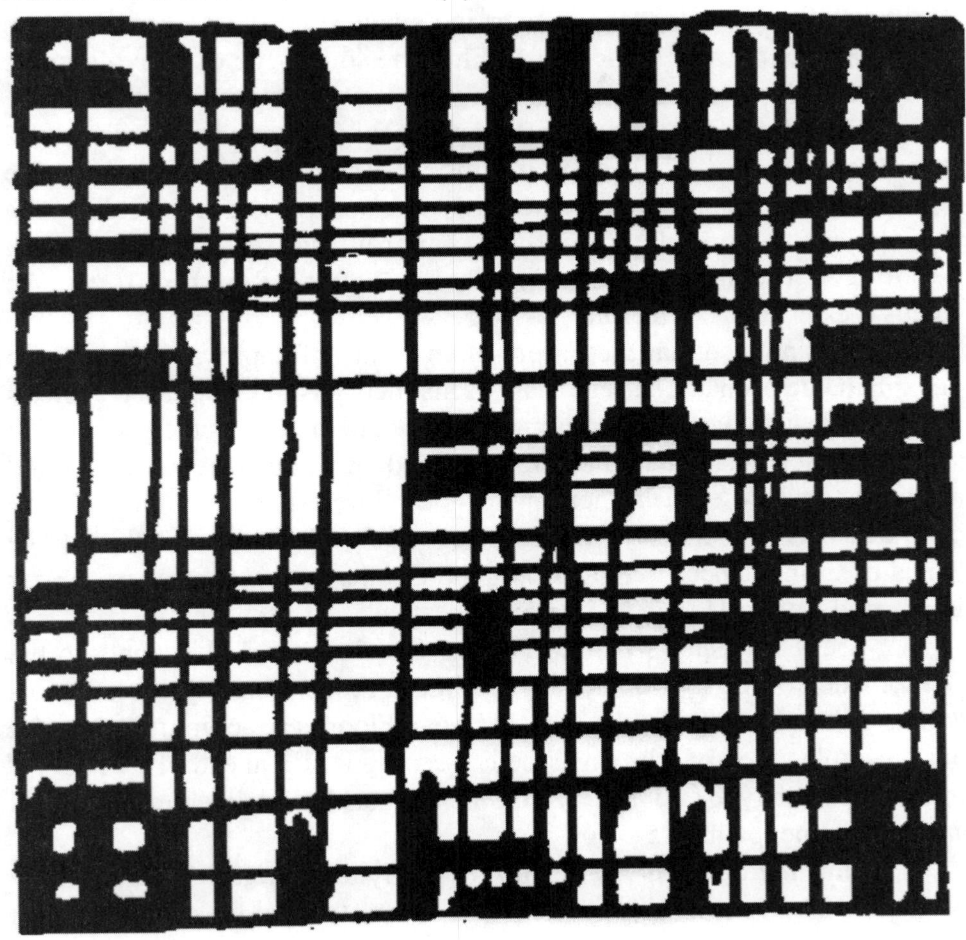

At face value, this sketch looks like nothing more than arbitrary lines and squares. To find its secret messages, you must look at it properly tilted. Only then do the squares elongate and lines shorten to create the words. This trick with perspective is possible because of the way we see. It works using almost any given name.

SYNCHRONICITY
(UNEXPLAINABLE COINCIDENCES)

A college professor once wrote me a note concerning the face on the Pacific Ocean Floor. He asked, "Could it not be some kind of (acausal) meaningful coincidence such as taught by Carl Jung?" He went on to ask if I knew about Jung's ideas concerning synchronicity. "Could this huge weeping face on the Pacific Ocean Floor be some type of unexplainable coincidence?"

Carl Jung, the eminent Swiss psychologist, used his invented term "synchronicity" to explain among other things, unexplainable coincidences. He said that the word synchronicity is used to "describe a meaningful coincidence without apparent cause (acausal)." Jung introduced the idea to America in his essay, "Synchronicity; An Acausal Connecting Principle" in 1955.[5]

Jung noticed that there were often plenty of unprovable answers to unexplainable events. The real heart of his theory is that supposed answers to unexplainable events were less important than the events themselves. He said, "Skepticism should, however, be leveled only at incorrect theories and not at facts which exist in their own right."[6] Interest in Carl Jung's theory exploded during the 1970's with writings such as Arthur Koestler's "The Roots of Coincidence."[7] Another is "The Challenge of Chance" by Alister Hardy and Robert Harvie and Arthur Koestler.[8]

Are there aspects of Carl Jung's "Synchronicity" that apply to the phenomenon of the face on the Pacific Ocean Floor? Possibly, but only to a certain point. The face on the Pacific Ocean Floor is a fact in its own right, but the explanation of why this face is weeping is still ahead of us. Also, if this is not some type of chance event, then it must be a deliberate act of God and not an unexplainable event.

Jung included in his definition of Synchronicity: "Extrasensory perception, psychokinesis, oracle consulting techniques, astrological horoscopes, prophetic dreams, deja vu, unconscious foreknowledge, omens and number series. It suggests an acausal connecting principle in nature, of equal importance to physical causality. 2: A meaningful coincidence, (syn*chro*nis'tic, adj; syn*chro*nis'ti*cal*ly,) adv."[9] If Synchronicity is a consideration, concerning the face on the Pacific Ocean Floor, then it is logical that one simple question should be addressed.

DELIBERATELY OR BY COINCIDENCE?

Has the face on the Pacific Ocean Floor been created deliberately or is it a coincidence? Is there a relating cause or not? There are so many relating factors (both scriptural and scientific) that answers may be coming soon. Conclusive evidence has not arrived, but what we know now can make even the most cautious researcher very excited. At least, now we have a basis for some valid study. The hypothesis is at least solidly becoming a theory.

Scientists are often fascinated by how we perceive certain optical illusions. Studies more than forty years ago found that what we perceive does not always directly correspond to reality. What we see is actually a subtle blend of the external world and the many lessons of our personal experience.[10]

Another factor is the effect of our habitually using perspective in our daily lives. Thus, ambiguities are sometimes difficult to reconcile when looking at two-dimensional pictures. In 1968 researcher Richard L. Gregory completed and published his study of visual illusions. He wrote that "Perhaps it's because the visual system has to make sense of a world in which everyday objects are normally distorted by perspective."[11]

How we look at pictures has everything to do with how we interpret what we see. This was apparent as I watched different people and their reactions when they were first shown the topographic map of the Pacific Ocean Floor. Even when comparing the map with our artistic overlay tracing, some people didn't see the face as well as others.

This was especially the case in the early 1980's when this discovery was new and had not yet received notoriety. How each individual reacted to this discovery was based on factors I had not initially considered. Christians and those of other faiths that believe in God were often quick to see the face and appreciate its clarity. I wonder if this is because people of faith are less afraid to accept supernatural phenomena.

Less convinced of the miraculous origin of the face on the Pacific Ocean Floor seem to be agnostics or atheists. Religious bias is not the only factor that affects individual interpretation of this image. The way someone studies or looks away from this map and its border tracing also influences his ultimate reaction. People who take a few minutes just looking at the topography of the Pacific Ocean Floor become more interested than those who do not.

SCANNERS

Scientific research has long since shown that individual eye movement, the way one sits or stands and even the rate of breathing, can often affect interpretation. In the early 1960's Ulric Neisser began his study on the cognitive operations involved in looking, seeing and recognition. He worked with pictures, numbers and word lists in his study. He referred to his test subjects as "Scanners."

One of his conclusions was that the Scanner must extract enough information from the elements of the context to make sure or at least to suggest that they lack the properties that define the object of the search. The Scanner first would recognize the context of the image or list studied. Neisser concluded that "there are evidently intermediate stages of perception; it is not a case of 'now you see it' or 'now you don't,' but of something in between." The time allotted for the Scanner to complete the cognitive operation was also a factor influencing recognition.[12]

Eye movement is also a factor in how we perceive. In 1968, E. Llewellyn Thomas found it fascinating how eye movement could influence interpretation. He designed a special camera to record where people look during such activities as driving and looking at pictures. His special "Eye Marker Camera" tracked and recorded the eye's glance. One of his studies recorded the eye movement of an X-ray search (of a man's lungs) made by a student radiologist looking for signs of pathology.

Another study recorded eye movements of a subject fixating on an inkblot. The subjects looked at Rorschach inkblots and wrote what the blots reminded them of. At first their eyes darted over the entire area of the blot, making many fixations. As time went on, the fixations became decidedly longer. Thomas concluded, "Perhaps these fixation times reflect a process in which the viewer, having realized that the blot offers little or no genuine information, is adding or generating meaning rather than merely accepting it." Thomas also noted, "It appears that the eye fixes on many things of which the viewer is not aware."[13]

It is common for someone to change his mind about what he perceives, when presented with a Rorschach test. It is easy to see why. The subject knows it is only an inkblot. This variable back and forth perception also becomes apparent to some first-time observers of the topographic map on the Pacific Ocean Floor.

Contrary to a Rorschach test, once someone clearly recognizes the face on the Pacific Ocean Floor, that is it. He is never able to look at the map again without seeing the gigantic weeping face.

MOIRE' PATTERNS

To further understand how the eye reacts to the interweaving of lines in their interpretation, other scientists have studied Moire' Patterns. One use of these studies was to examine biological specimens that have refractive indexes close to that of water. This principal also provided a quick and easy means for testing lenses.[14]

Moire' effects are visual patterns which occur when one pattern with "open and blocked areas" is placed on another. The most common moires are made by repeated systems of lines or dots, concentric circles, radiating lines or spirals.

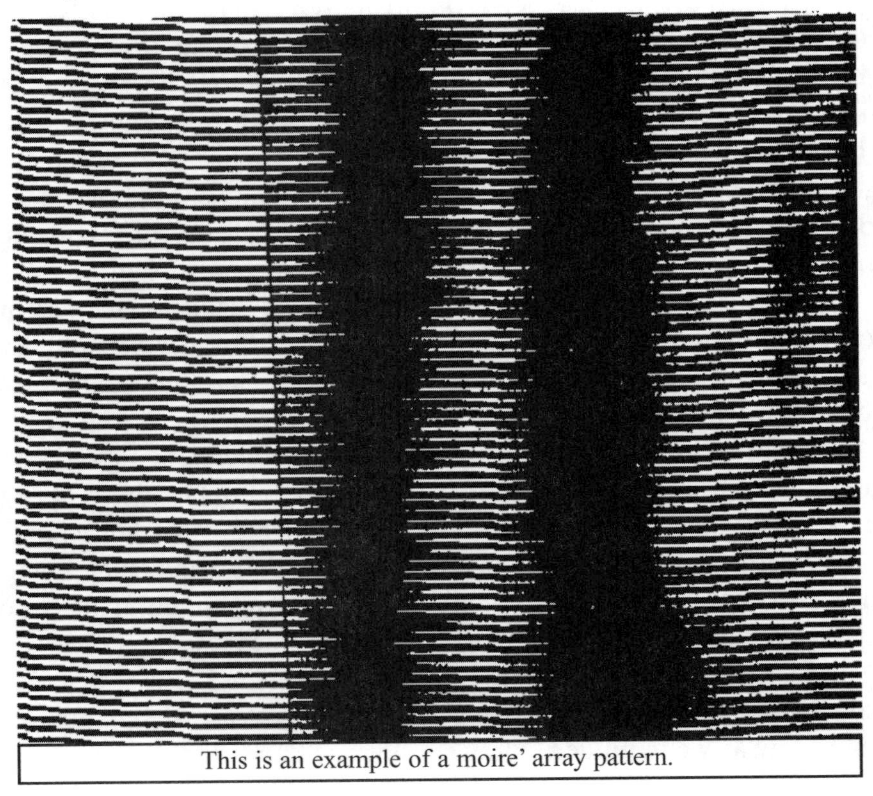

This is an example of a moire' array pattern.

INFANT VISUAL PERCEPTION

In 1966, T.G.R. Bower completed his study on how infants perceive pictures and three-dimensional objects. He wondered whether infants naturally see things in the same general way as adults or whether they must learn to do so. To learn the answer, Bower conducted an entire series of conditioned experiments with infants.

He found that infants learned by responding to real shape and not the retinal shape (like those in photographs, maps and slides). The infants in these studies were obviously able to register variables containing complex information. Bower was fascinated as he wrote, "Yet they were unable to use the information given in slide presentations. This failure must surely mean that infants are not sensitive to the kind of information that can be frozen on the plane of a picture."[15]

VISUAL INTERPRETATION ALWAYS DOMINATES

Another important factor influencing our interpretation of what we see is the sensation of touch. Irvin Rock and Charles S. Harris conducted experiments in 1967 to find out how humans learn to see. They wrote, "It has been held that *human beings must learn how to see,* and that they are taught by the sense of touch. New experiments demonstrate however, that vision completely dominates touch and even shapes it." This study concluded, when touch contradicted with sight, the visual interpretation always dominates. The conclusion of the team's work resulted in this written statement which contradicted previously popular theories: *"There is no convincing evidence for the time-honored theory that touch educates vision and that there is strong evidence for the contrary theory."*[16]

Scientific research has resulted in a better understanding of how the right and left eye attempt to dominate the other. This binocular rivalry can occur when the stimuli from one eye may be dominant with a corresponding suppression of the stimuli from the other eye. The dominance may fluctuate from eye to eye, continually switching.[17] Which eye is dominant is not a constant. Sometimes it is the right eye and sometimes it is the left.

How we look at things and the habit patterns in the way that we see are key factors. We do not know if any ancient civilizations understood optics the same way as we in modern times. Historians know that there has always been considerable involvement with looking at things properly. The requirements included a quiet heart and a still body. Another was a technique of gradually crossing the eyes. This was not done like a comedian who might cross his eyes as a joke. It was more like one gazing at a single thing while in deep thought. This technique resembles what is necessary to see images in modern computerized stereograms.

Many ancient societies found it desirable to master a deep trance state while the eyes were opened or closed. The Mayans fastened small wooden balls to their children's heads so that these balls dangled in front of the eyes, causing the children to become cross-eyed.[18] Although they presently live in the same world as modern industrial nations, the Wodaabe tribe in Africa practice what they believe to be ancient Egyptian beliefs. Men, women and children cross their eyes all day long. This tribe believes they are direct descendants of the ancient Egyptians. They hold that only by keeping their eyes crossed can they see the world as it really is.[19]

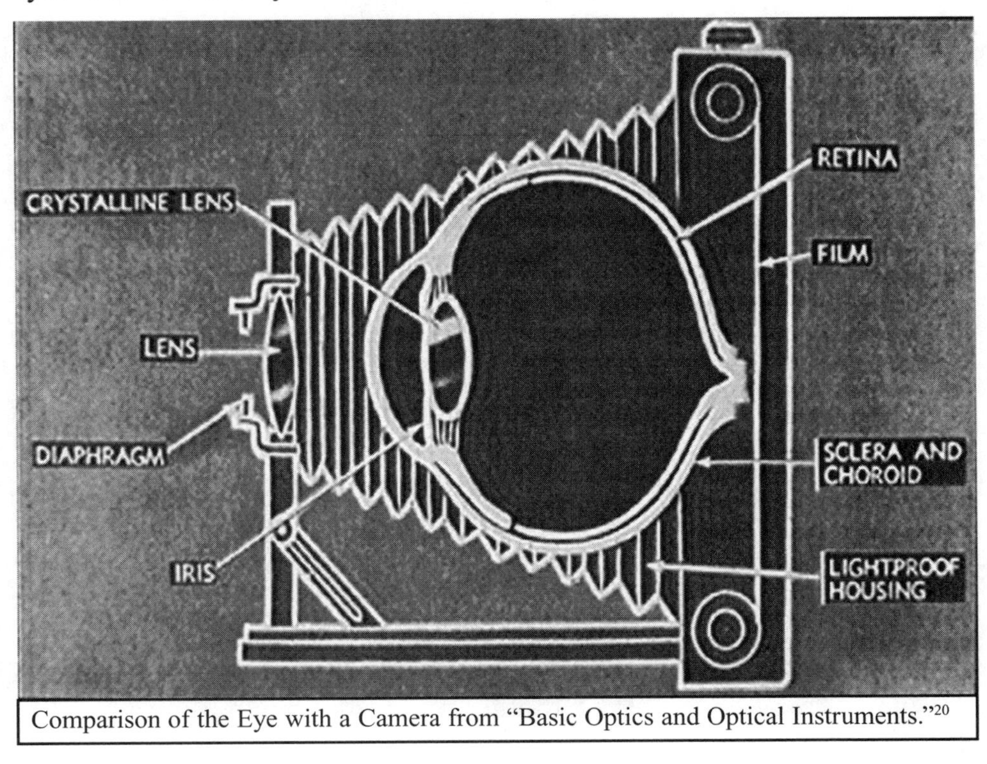

Comparison of the Eye with a Camera from "Basic Optics and Optical Instruments."[20]

To the Ancient Egyptians, the Eye was the Seat of the Soul.

The ancient Egyptians believed that the eye was the seat of the soul and all powerful. They concluded that the eye possessed the real means of deliverance from their imprisonment in matter.

Through the crossing of the nerve fibers in the optic chiasma, the left half of the brain sees with the right eye, and the right half of the brain sees with the left eye. This interweaving of perception renders us conscious of objects and their forms. To the ancient Egyptians the eye related to the divinities of weaving or crisscrossing, which they called "sia." This also was a link to their divinity of knowledge, whose name is written in hieroglyphics as a cross-weaved, square patch of cloth. This is because the eye, our predominant means of gaining knowledge, shares the principle of crossing by which a weaved fabric also relies.[21]

Considering this, the Wodaabes might be correct in their thinking. Maybe one way to really see things in a proper manner is to practice crossing our eyes.

SEEING VISIONS AND DREAMING DREAMS

Some scholars believe that the ancient Hebrews may have developed a technique of self-hypnosis that gave them the experience of ascending into Heaven and joining God on his throne. Others hold to the idea that all such manifestations are intrinsically an act of God. The approval of this or similar techniques can be seen elsewhere. One need only look to the teachings of the ancient prophet Joel (v28) ". . . your old men shall dream dreams, your young men shall see visions."

Chinese ideas relating to how one looks at things may seem similar to the Mayans or Hebrews but the Chinese goal was not necessarily seeing a vision or dreaming a dream. Their goals were more down to earth. To the ancient Chinese, the conservative approach held much more appeal. Their focus was more toward recognizing the reality in art, mathematics and nature. In the second century B.C., a Chinese naturalist completed the book "Huai-nan tzu."

THE POWER OF NON-VERBAL COMMUNICATION

Among the students of this book was Wang Chung (A.D. 27-96). His teachings involved a method of examining actual objects and events. His naturalist ideas even extended to Heaven and Earth, which he considered as natural objects without will. He also taught that hallucinations were there because of sickness.[22]

While producing my documentary, "Scientific Proof of the Deliberate Supernatural,"[23] I felt it would be helpful to consult with an expert in the field of non-verbal communication. Maybe a specialist in this area could help us better understand some unique things about the quality of non-verbal communication. I was curious how pictures, like those on our ocean's floor, might be an excellent way for God to communicate with humanity.

Our interviewer, Michael Ellington questioned Mr. Scott Rodriguez, who at the time worked with the Communications Department at California State University at Los Angeles. Mr. Rodriguez received national honors for his abilities in speech communications. Since he worked as an instructor in this field, I thought he could add insight to our questions. His expertise covered a wide range of speech areas that included non-verbal communication. The following is some of that interview.

A PICTURE *IS* WORTH A THOUSAND WORDS

Ellington:

Mr. Rodriguez, is a picture worth a thousand words?

Rodriguez:

Well, as long as you're not speaking in the technical sense of theory that is documented by communication experts, I think that it's generally possible for people like us to talk about a picture being worth a thousand words. So, by looking at pictures and how they translate into non-verbal communication, I think you probably could say that if not a thousand words, a lot of words at least ought to be attributed to any picture.

Ellington:

In relation to this map of the Pacific Ocean Floor and its accompanying border tracings, what might the image relate?

Rodriguez:

Well, the actual image that is traced, looks like a man with a track of tears going down his face.

Ellington:

Where would the eye be on this map if there was a tear-track going down the side of the face?

Rodriguez:

Well, logically the eye would begin right about here (pointing to the coast off Manzanillo, Mexico).This looks to be the front of the face, so if I were to read where the eye came out, it would, I guess, be right about here.

Ellington:

As an expert in all levels of communication, if you wanted to relay a simple message about good and evil, which would communicate to all ages and languages, what would be best, words or pictures?

Rodriguez:

Cognitively again, and this has been proven by psychiatrists, words and images do not have the same effect on people. Words are bound by culture and language and connotation and other aspects that are not tangible. Whereas images are not bound to particular languages or even particular cultures. They even transcend age groups. You can touch people whose ages are lower than the age of reading with these kinds of images.

Ellington:

What I'm looking for now is not so much your expert opinion, as an expert in communications, *but more your personal gut-level-reaction*. What do you think about Lloyd Carpenter discovering these kinds of images on these maps?

Rodriguez:

Well, I think it's just crazy. The man is some sort of a wild genius, I think, to have gone and made these attributions on the basis of evidence which is actually physical and here on the planet. And in order to make attributions to an image, generated, not by an artist but by the world itself, I would think shows some sort of plan or design in the universe. So, if indeed these kinds of attributions can be made using the geographical, the geological structure of the earth, then there must be something about the geological structure of the earth that we don't yet understand.

Ellington:

Back to the map of the Face on the Pacific Ocean Floor (he points to the eye area of the map). Are you curious about where this face appears to be looking?

Rodriguez:

Well, uh, yes as a matter of fact. Without all the relating data, it's very hard to tell where the eye might be looking to. So yes, actually that is quite interesting, isn't it?

Chapter 5

THE ANATOMY
OF THE
PACIFIC OCEAN FLOOR

PERSONAL IMPLICATIONS FOR EVERYONE

It is one thing to say that the topography on the Pacific Ocean Floor looks like the face of a weeping man. This is not a matter of much debate. We see that the shape of the underwater mountain ranges (on a map of the Pacific Ocean Floor) does appear to be the face of a weeping man. We see the right side of his head, neck and face. We see substantial detail in the eye area. He looks upward, as his hair flows behind him. But at some point, serious researchers begin asking this question: *"How many substantiating features, supporting the likeness of a man, are there?"*

Why is this such an important question? Because, when we begin comparing features on the map with those of a man, an obvious follow-up question is the result. At what point in logical thinking does the possibility of this being a coincidence cease? Locating the eye, nose, chin, mouth and hair is simple. It is the very simplicity in finding these features that makes it easy for the observer to assume they must be coincidental. If not, then this phenomena has some personal implications for everyone

GETTING PAST COINCIDENCE

Believing that the face on the Pacific Ocean Floor is simply some sort of Synchronicity or unexplainable coincidence, is safe thinking. There is a strong drive in some individuals that causes them to prefer denying the reality of the face on the Pacific Ocean Floor. If this big face is a coincidence, then we can think anything we want to about it. We can hold it up, or we can simply discount it as just one more silly, unexplainable thing. If we believe that this face is just a coincidence, we can make jokes about it, or curse it, or ridicule it, or just forget it.

If we compile enough evidence to conclude that the face on the Pacific Ocean Floor is not a coincidence, then how we think about it will diametrically change. If this image is not a coincidence, then it must be a deliberate creation from God. If not, then we must ask who put the face on the Pacific Ocean Floor and why? If some other person or persons did put it there, then they are obviously in control of an incredible ability to communicate with an entire planet at once.

FROM HIS EYEBALL TO HIS BRAIN

Thinking of God as one who would place his imprint and identity on our planet is not new. Throughout history, philosophers have seen the Earth as synonymous with the anatomy of the Creator. My files include stories and myths that range from the Earth being God's eyeball, to it being his brain.

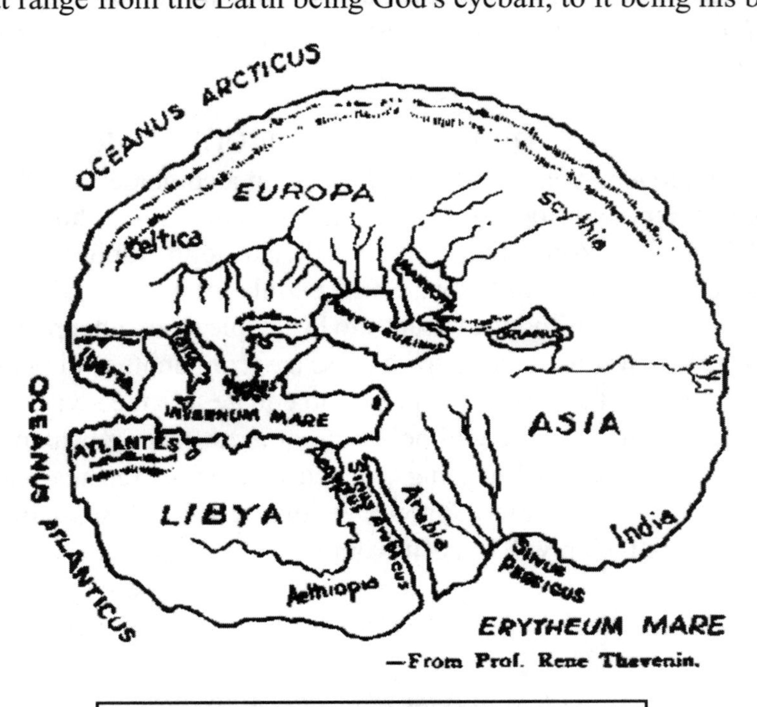

Map of the World from the Time of Herodotus

People who believe in God and go to the Bible for back-up will have fewer problems with this discovery than those who do not. Some Christians, Jews and Muslims, because they each look to their versions of the Bible, may find themselves more accepting than others. Some religions could change in a dramatic way, but only time will prove this conclusively.

No one knows what will happen when the leaders of some faiths become convinced that the face on the Pacific Ocean Floor is not coincidental. It is only natural that some leaders will conclude that this face was put there deliberately by God. With that, some may then say that not only was this face put there by God, but that it is also a *representation* of God's own face. To convince followers of their respective faiths, some teachers may point to details in the Bible and on topographic maps to substantiate their various religious positions. What all this really means is that science and religion are uniting to form a theory. This study is at the foundation of the theory we speak of, but a foundation is not a building. If religion proceeds with caution and looks to science, then everyone will benefit.

THE DILEMMA OF THE DOUBTER

People who believe that this face is not a coincidence, but also don't believe in a personal God, could have their own dilemma. Those more prone toward New Age, Agnostic, or even Atheistic philosophies, may at first hypothesize all manner of things. Some people might guess that this face is something put here millions of years ago by people from other planets. Others might figure that this is somehow a message from another dimension.

To those of any belief system who equate this face with the "End of the World," the near future will be a time of great fear and calamity. To see prophecy about the Earth tipping over on its axis, or the sea becoming as blood is enough to frighten anyone. Fortunately, those who are secure in their eternal salvation will benefit from uncanny strength.

Psalms 46:2
"Therefore will not we fear, though the earth be removed, and though the mountains be carried into the midst of the sea;"

There is much less fear when a person knows that nothing on this earth can take away his eternal security. Some people might look at some of this as blind faith, but it is not. God has a Holy Spirit that is able to give a certain peace that passes all understanding.

THE FOCUS IS ON THE ANATOMY

At the same time as people are discussing the merits of the face on the Pacific Ocean Floor, as it relates to Bible prophecy, a second area will gain equal interest. This will be a detailed study of the face on the Pacific Ocean Floor, with *anatomy* as its focus. At one end of the spectrum will be level- headed researchers using similar identification methods to those used by law enforcement agencies.

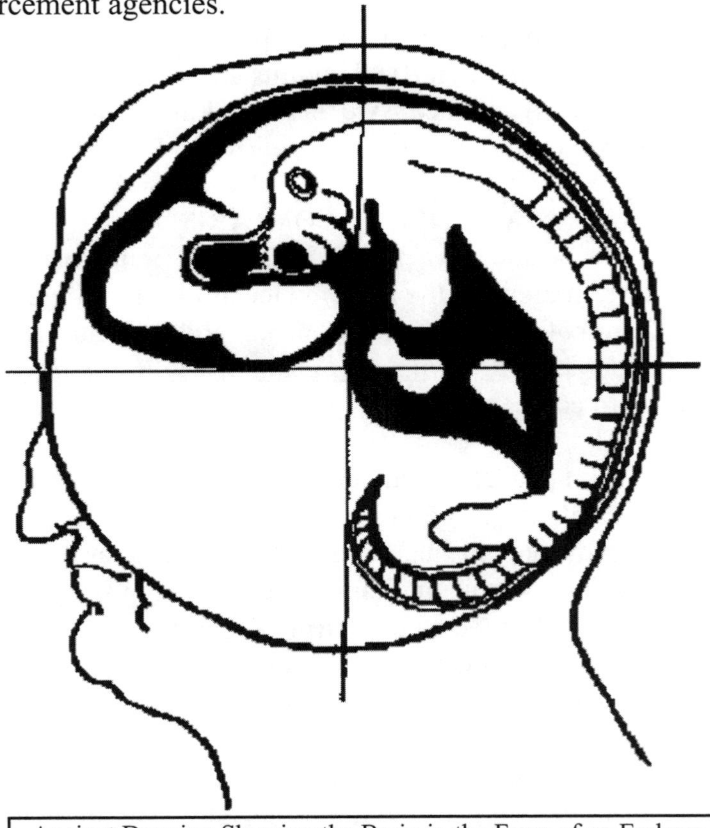

Ancient Drawing Showing the Brain in the Form of an Embryo

Pertinent evidence will be compiled in accordance with strict scientific procedures. At the other end will be the more esoteric researchers who might use various religious and occult methods. These could range from a hunt for the shape of the human body in our solar system, to finding the likeness of a human fetus mirrored in the brain.

HOW MANY IDENTIFIABLE POINTS ARE REQUIRED FOR POSITIVE PROOF?

There are only six areas of a fingerprint that are differentiated when identifying a fingerprint of a person. They are the "double loop, the whorl, the central pocket loop, the arch and the tented arch."[1] The reason I mention fingerprint identification is to show that there are valid methods for identifying people and things. With this principle in mind, we might wonder how many matching points can be found concerning the face on the Pacific Ocean Floor. How many identifiable points are needed to prove that which looks like a face, really is a face? Are there any kind of similar guidelines we can follow to keep our points of reference verifiable?

A GENERAL COMPARISON

The chart on the next page is to show a comparison between the location of certain areas of the Pacific Ocean Floor and the human head. This cross section of the human head is from "Fyfe's Anatomy" and is being used for a good reason. *It displays a human head divided down the middle.* This allows us to compare the inside of the head in a manner not otherwise possible.

On the next page, the numbered circles next to the human head and the map of the Pacific Ocean Floor are not arbitrary. They are meant to show specific areas on the human anatomy and the related areas on the map. These lines are utilized as a simple guideline and are in no way comprehensive. It will take a far more detailed study to truly decipher matching localities and their more intricate relationship to each other. Maybe someday this can be done in a truly scientific manner.

TRANSPARENT LIKE A SPIRIT

With a careful study of the Rand McNally map of the Pacific Ocean Floor, you might notice something unusual. *Many of the identifiable features found on the face on the Pacific Ocean Floor are actually located inside the head and are not normally visible in humans without the benefit of X-ray or ultrasound.* In many ways, much of the big face on the Pacific Ocean Floor is almost transparent *like a spirit.* An example is the area off the coast of Peru, showing *teeth* inside his closed mouth. Another is the *brain stem* comparison with the Fiji Plateau and Tonga Trench.

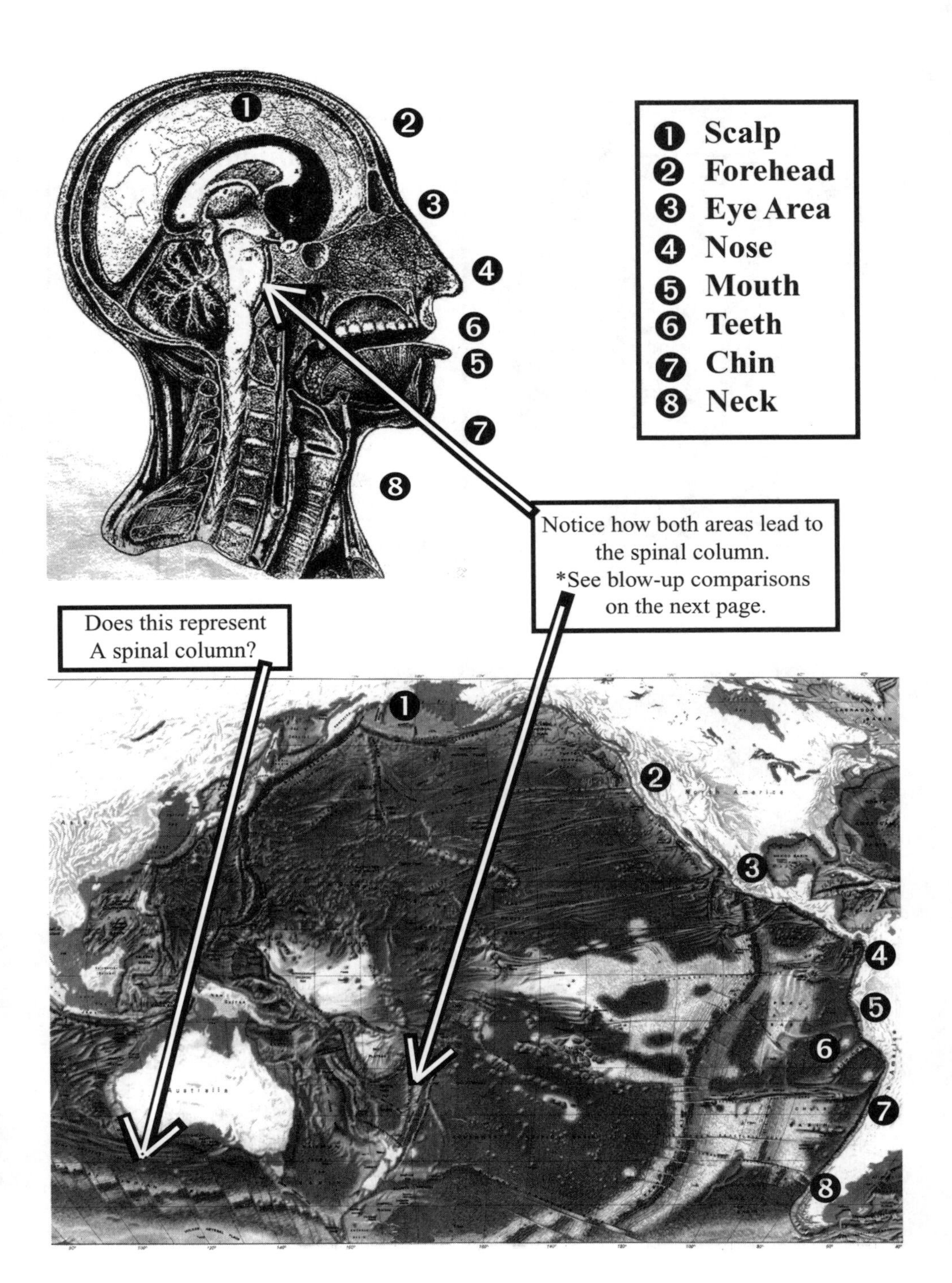

❶ Scalp
❷ Forehead
❸ Eye Area
❹ Nose
❺ Mouth
❻ Teeth
❼ Chin
❽ Neck
Notice how both areas lead to the spinal column.
*See blow-up comparisons on the next page.
Does this represent A spinal column?
North America
Australia

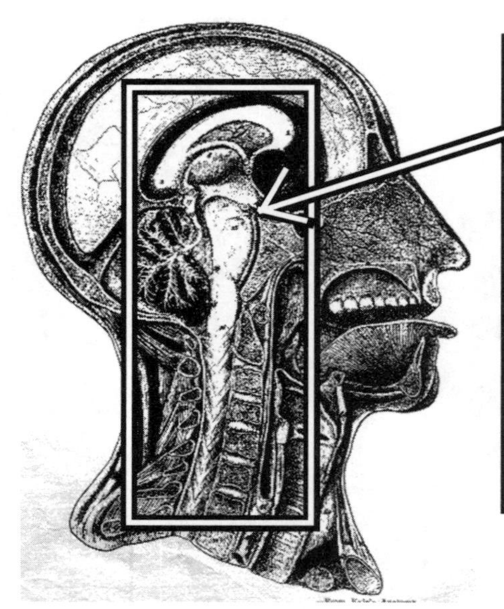

A Comparison
THE PITUITARY BODY
and
PINEAL GLAND
with the
FIJI PLATEAU,
KERMADEC TRENCH
and the
TONGA TRENCH

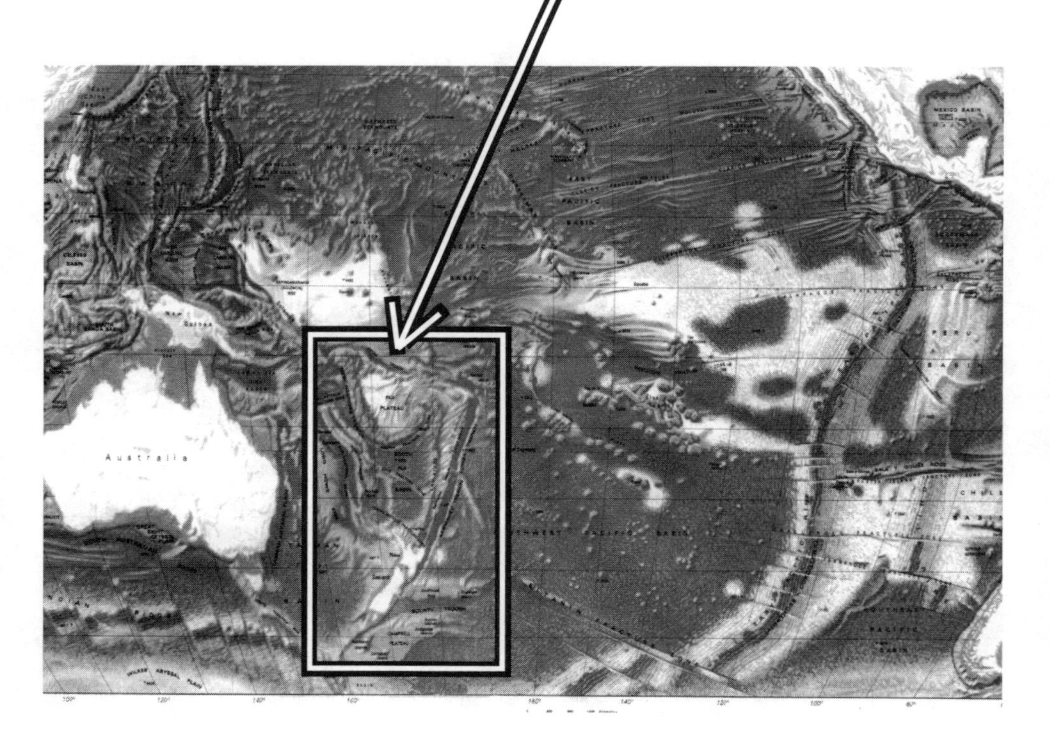

???????

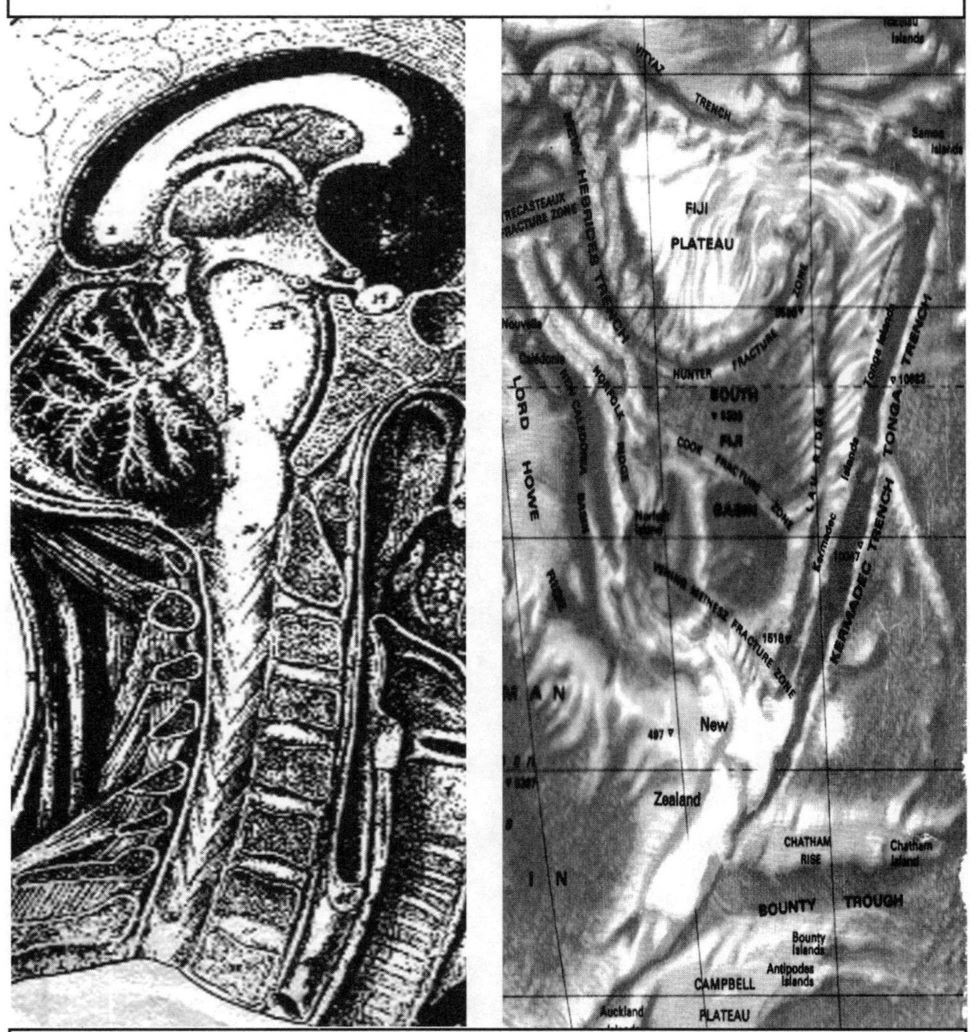

This is a side by side comparison of the interior of a human head from the previous page, with the corresponding area on the Pacific Ocean Floor map. The proximity and angle are slightly different, but various visible similarities to the pituitary and pineal gland, leading to the brain-stem and spine are evident.

THIS INSET SHOWS THE SECTION HIGHLIGHTED ON THE NEXT PAGE.

Face section of Rand McNally Map of the Pacific Ocean Floor.©

THE NAZCA RIDGE

Could This Represent Teeth?

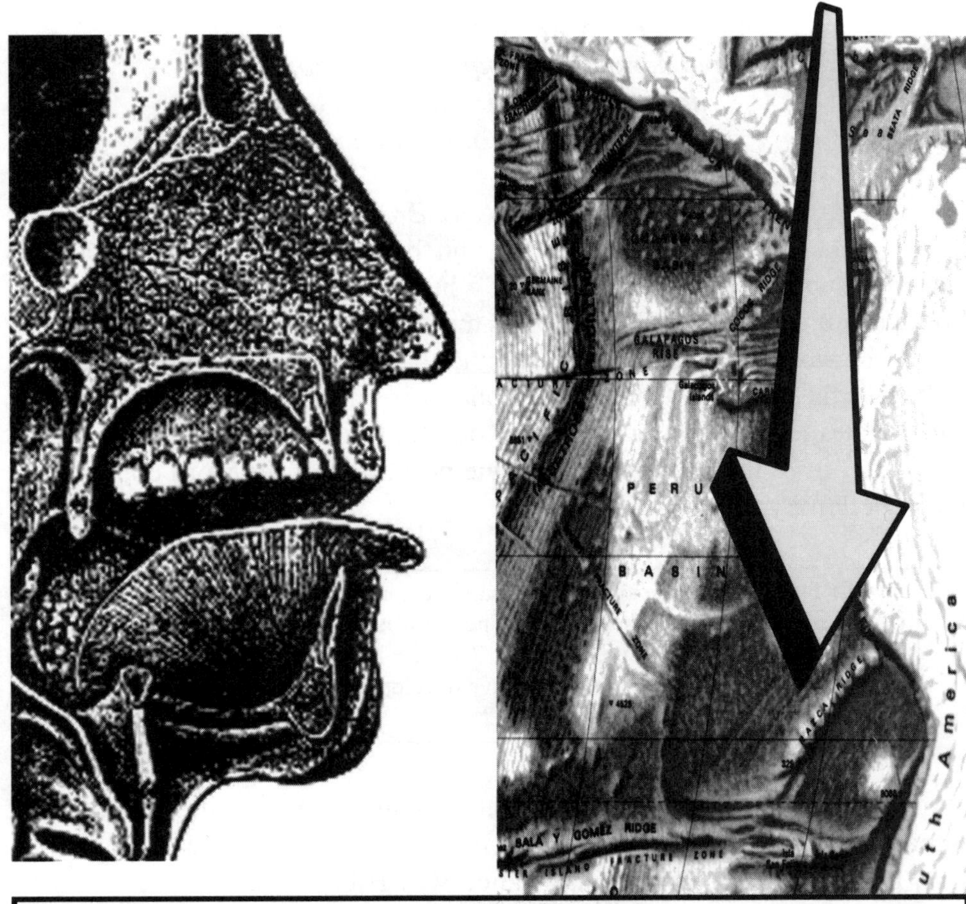

This is a side by side comparison of the interior of the human head (from Fyfe's Anatomy), with its corresponding area on the Pacific Ocean Floor map. Here, the focus is the Nazca Ridge, which is located just south of the Peru-Chile Trench. Could this underwater mountain range be representative of his teeth? Take a moment to compare the images on this and the previous pages. Why does this image seem to also show the inside of his head? Notice the similarity in the direction of muscle structure inside the mouth and leading to his neck.

Chapter 6

THE SIGNIFICANCE OF GOD'S FACE

Genesis 1:2
"And the earth was without form, and void; and darkness was upon the face of the deep. And the spirit of God moved upon the face of the waters."

THE APPEARANCE OF GOD

Whenever I use the phrase "face on the Pacific Ocean Floor," it always refers to substantially more than just a human-looking face. In a way, this nomenclature applies in general, to its more ancient meanings. Originally, in Latin, the word "face" meant "appearance." So in a very real way, whenever I refer to the "face of God," the emphasis is on the "appearance of God." This relates to the original terms, not only in Latin, but also how they were translated from the Hebrew script. One definition of the word comes from the "Dictionary of Word Origins."

> **FACE:** Through Fr. face and Prov. fassa, face comes from Latin, facies, which first meant appearance, then visage. The word is from the Latin, "fac," as in facem, torch; hence, to shine, to appear. The face is that by which a thing shows itself: from this, the other meanings have developed, from facing the enemy to behaving as to lose face.[1]

While I was still in college, endorsements of encouragement were important to me. So I sought counsel from every type of expert imaginable. One person of considerable assistance was Bible scholar Rabbi Alan R. Lachtman. He told me that the Hebrew word for "face" can have a double meaning. Not only does Genesis 1:2 mean the surface of the waters, but also, it can refer under certain circumstances to a human face. At the top of this page we see that the second verse in the Bible mentions the word "face" twice. After examining my map comparisons, Rabbi Lachtman wrote a letter of encouragement. A copy of his letter is on the next page.

TO WHOM IT MAY CONCERN

Not only did the Rabbi clarify the Biblical term "The face of the deep" (Gen. 1:1-2), but he even went an important step further. In the fourth paragraph he adds, "The face of the deep, may in fact be the discovery of the sonogram revealed to me by Lloyd Carpenter."

temple beth david
of the san gabriel valley
בית דוד

Alan R. Lachtman, Rabbi
Irwin Frazin, President

December 7th, 1982

TO WHOM IT MAY CONCERN:

Re: THE LLOYD CARPENTER STUDY

I have personally viewed the research and investigation by Lloyd Carpenter regarding the appearance of the Pacific Ocean Mountain Range through a sonar study, which reveals the likeness of a human head which contains similarities to the anatomical features of a human-like head.

In my opinion, this finding deserves further study because of the merit of the initial findings.

I find this most interesting when the reading of the genesis narative, in the first Chapter, second verse, speaks of "the face of the deep" על פני תהום.

It appears that the "eye" is located near Guadalajara and seems to point towards Israel. "The face of the deep" may, in fact, be the discovery of the sonagram revealed to me by Lloyd Carpenter.

I find equally amazing, the configuration of a dragon in the sonogram of the North Atlantic Ocean, which echoes the Mesopotamian tradition that the earth and its heaven were formed with the aid of the sea dragon, Tiamat.

Sincerely yours,

RABBI ALAN R. LACHTMAN

THE ANCIENT OF DAYS

In the seventh chapter of the book of Daniel, we can read about the physical image of God three times. Each time Daniel uses the term "Ancient of Days," he describes the judge of all humanity. Each reference occurs during his description of four passing kingdoms that he calls the "four beasts." These fallen nations make way for the kingdom of God. Daniel had dreams and visions while sleeping in his bed. It happened during the first year of the reign of Belshazzar, king of Babylon. The prophet, upon awakening, wrote down everything he saw while he dreamed in great detail. Then Daniel started talking about all the things that he had seen in his dream (Dan. 1:1). After seeing the fall of four great kingdoms, *Daniel sees the face* of what he first describes as *"a man."*

We know that Daniel sees the face of God because he elaborates and goes on to tell his dramatic story in great detail. Daniel recognizes God's eyes and mouth and hears a voice from this mouth that is "speaking great things." Next, Daniel has a full view of God on his heavenly throne. Daniel describes the hair on God's head (like wool) and the garment (white) he is wearing. He even expounds on the reflective brightness (like burning fire) surrounding God's throne.

Daniel 7:9
"I beheld till the thrones were cast down, and the Ancient of days did sit, whose garment was white as snow, and the hair of his head like the pure wool: his throne was like the fiery flame, and his wheels as burning fire."

The second time Daniel mentions the Ancient of Days is just after he explains that the lives of the four nations are preserved in a much weaker state under God's new Kingdom. Then, with help from "one like the Son of man," Daniel is brought before the throne of God.

Daniel 7:13
"I saw in the night visions, and, behold, one like the Son of man came with the clouds of heaven and came to the Ancient of days, and they brought him near before him."

From Montfaucon's Antiquites

It was in this form, that Jehovah was generally pictured by the ancient Hebrew mathematicians, Bible scholars and Kabbalists. In this form, God was called (by the Greeks and Gnostics) the "Immortal Mortal." The Hebrews referred to him as, "IHUH."[2]

Then Daniel goes on with the recollections of his visions. He is worried and anxious. His dreams begin to trouble him (v15). It is no wonder. Daniel sees the terrible "end times" war with the "Saints of the Most High," being defeated (v21). Then he sees that they are rescued by God, in person, as he (God) assumes the role of the savior of the world and the Ancient of Days. It is also during this time that God saves the world from ruin.

Daniel 7:22
"Until the Ancient of days came, and judgment was given to the saints of the most High; and the time came that the saints possessed the kingdom."

In an effort to set forth an appropriate figure for the Christian doctrine of the Trinity, it was necessary to devise an image in which the three persons, Father, Son, and Holy Ghost, were separate and yet one. Figures similar to the one above, wherein three faces are united in one head, can be seen throughout different parts of Europe. Scholars consider this a legitimate method of symbolism.[3]

A MODERN MIRACLE WITH ANCIENT ORIGINS

The images on our ocean's floor tell a complete story of "Good versus Evil." Each figure is huge, and together they cover about 70% of our planet. These pictures create a story that fits the central theme of the Holy Bible. One figure (the face on the Pacific Ocean Floor) is so enormous that it covers almost one-half of this planet. It alone is a modern scientific discovery with ancient and holy origins. Can the study of ancient scripture help us better understand why this face exists?

GOD'S FACE IN THE BIBLE

I can think of no better way to learn about God than by reading the Holy Bible. Throughout the scriptures we can find references to God's appearance and nature. Could the face on the topography of the Pacific Ocean Floor depict God's weeping face? Could such mercy on his brow be a clue? The tear-track seen along the right side of his face is also part of the largest mountain range in the world. It is called the "East Pacific Rise." The various recognizable themes in this study link to ancient scripture more than we might expect. That is why this phenomenon stimulates curiosity about the face of God. Whenever anyone is curious about the nature of God, the Bible can come in very handy.

Researching Bible verses that might relate to the figures on the ocean floor requires scrutiny. Some of the references apply specifically while others indirectly. Still, it is a simple word study that we can appreciate. How important is the *face* of God in scripture? Can these passages tell us more about God? There are some subtle clues in the first verses of Genesis.

GOD'S SPIRIT MAKES A PHYSICAL IMPRINT

We see the word "face" referred to twice in the beginning passages of Genesis. First we learn that darkness was upon the *"face of the deep."* Then we see that the Spirit of God moved upon the *"face of the waters."* In both cases the word "face" has a dual meaning. The Hebrew term, used here, for the word "face" is "Paneh," (Pronounced pawneh'). First it can mean the surface of the waters or the surface of the ocean floor. Secondly, this word is also used to describe <u>a man's face</u>.[4]

These verses could be describing the surface of the waters <u>or</u> the surface of the topography of the underwater mountain ranges. But why the dual meaning also referring to a human face? Those who appreciate the subtleties of Hebrew writing might see that we could interpret the beginning of the Bible this way: "In the beginning, God's spirit created a human appearing face or something that looked like a face." Could the dual meaning also suggest the idea that somehow the *surface* of the waters also looked like, or reflected the appearance of a human type face? Did this face at the bottom of the sea become real when God's spirit came upon the surface of the waters? The verse simply reads, (Gen. 1:2) *"And the Spirit of God moved upon the face of the waters."*

If this passage does refer to the "face" of God, or the "face" (of the Holy Spirit of God), then what? Should we take a closer look at the last great promise in the book of Revelation? Scholars often interpret this prophetic verse as apocalyptic in scope. The prediction is that during the period known as the "End of the World," all humanity will physically *see the face of God.* The fact that *"his name shall be in their foreheads"* is a mystery that for the time being remains unanswered.

Revelation 22:4
"And they shall see his face; and his name shall be in their foreheads."

The first events and last great promise in the Bible highlight the face of God. Today, we can look at a modern topographic map of the Pacific Ocean Floor and see the face of a weeping man. If this face does represent our Creator God, does it also fulfill the prophecy of Revelation 22:4? Is this scripture even more evidence as to why the face on the Pacific Ocean Floor might be a harbinger to the apocalypse?

SOME BASIC QUESTIONS

Whether these Bible verses also relate to the face on the Pacific Ocean Floor is difficult to say. Still, this kind of study can lead to some provocative questions. One key question might be this: If the face on the Pacific Ocean Floor really does represent the God of the Bible, what should we do about it? Also, we might ask what is the best way to investigate and what other clues do the scriptures hold?

THE RIGHT SIDE

First-time observers of the face on Pacific Ocean Floor soon get past convincing themselves that it is in fact a reality. Then they reach another level, noticing certain peculiarities. One is this: All we really see is the *right side* of the head and various details inside the head. This is significant when we realize that in ancient literature, the right equates with light and goodness. The left of everything was to ancient man only a mere shadow of the truthful and honest right hand.

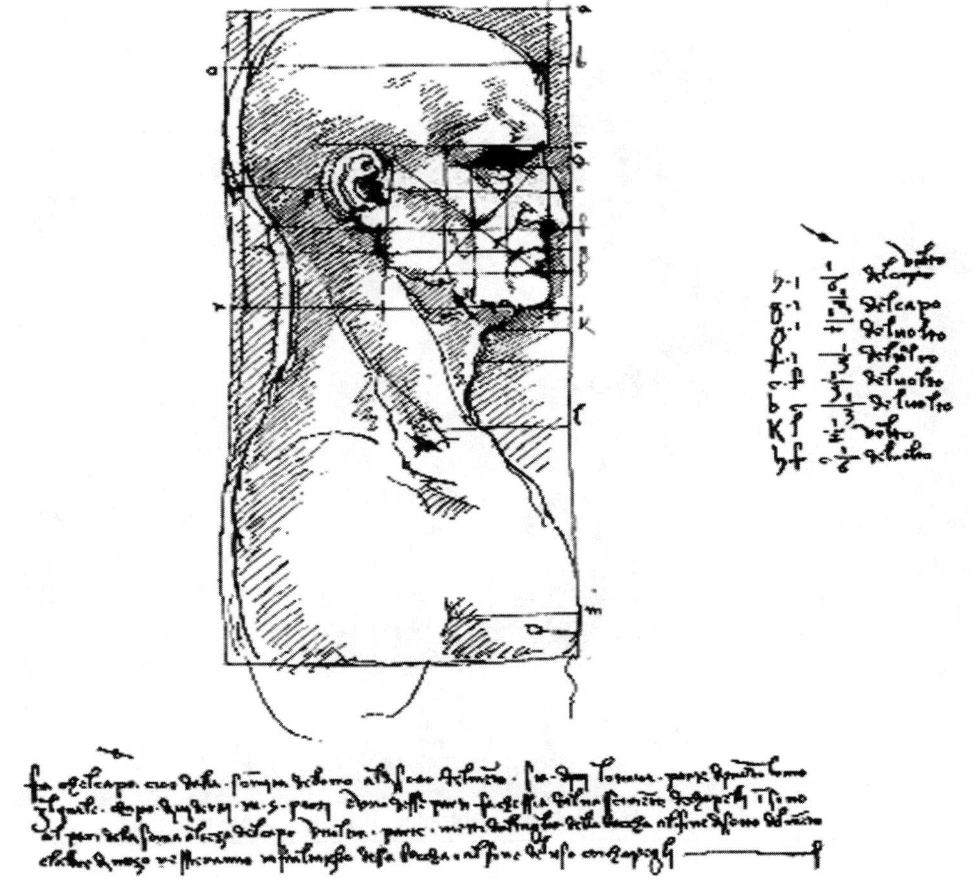

From Del Cenacolo Di Leonardo Da Vinci Di Giuseppe Bossi.

Leonardo Da Vinci's Canon for proportions of The Human Head[5]

In symbolism, the body divides vertically into halves. The right half holds light and the left only darkness. In ancient times men fought with their right arm and defended the vital centers with their left arm, carrying a shield to protect them. Also, in many cultures the right half of the body was considered masculine and the left half feminine. Just as goodness and Godly prayer is the correct and proper way to be, black magic has always been known as the left handed path.

Horus the Hawk stands guard at the entrance of Edfu Temple in Egypt.

THE FACE OF GOD IN ANCIENT EGYPT

In Egyptology, the "Hawk God" Horus is famed as a symbol of the resurrection from the dead. His name was synonymous with the rising Sun. As the son of Osiris, they believed he was the "Son of the Living God." Horus went by various other names also. Often the texts refer to "Horus of the Horizon." The ancient Egyptians also relished using dual meanings with words and phrases.

They wrote the word "Horus" using the hieroglyphic designation of "Heru." The second meaning of the word "Heru" was "face," such as a man's face. Therefore, it is also an accurate translation to interchange "Horus of the Horizon" to mean "Face of the Horizon."[6] Horus to the Egyptians was a deity and the face of their deity was synonymous with his name.

DOES GOD ASK US TO SEEK HIS FACE?

When we examine the emotion expressed by the weeping face on the Pacific Ocean Floor, we notice something else. He is looking at something that causes him grief. His gaze is direct and focused. What can it be? What is the object of his affection? The Bible teaches that God wants to be, not only acknowledged, but also worshipped. In response, God makes certain promises. The Bible tells us that seeking God's face involves a process that includes repentance and prayer from his people leading to forgiveness and healing from God. We can see an example of this in:

II Chronicles 7:14
"If my people, which are called by my name,
shall humble themselves, and pray, and
seek my face, and turn from their
wicked ways; then will I hear
from heaven, and will
forgive their sin, and
will heal their land."

BIG MAN - LITTLE GOD?

The Bible describes man as someone created in the image of God (Gen. 1:26). A natural outgrowth of this was the fabrication of a secret theological system in which God was considered the Grand Man and conversely, man as the little god. Continuing this analogy, the universe is regarded as a man and, conversely, man as a miniature universe.[7] This is not to say that man is a god or that God is a man. One argument against this fallacy is that man has a built in need or inclination to worship his creator. If this were not so, religion would not be a reality.

NEVER WORSHIP AN IMAGE
EVEN AT THE BOTTOM OF THE SEA

The idea that an image of any kind would bring us closer to God is repugnant to Bible believers. No picture of any kind can match the full totality of God. We see the validity of this in the commandment (Ex. 20:4), *"Thou shalt not make onto thee any graven image,..."*

As one of the "Ten Commandments," this section of the Law goes on to condemn those who would worship an image at the bottom of the sea. Below, we put that part in bold text for emphasis. The initiates of old warned their disciples that an image is not a reality but merely the objectification of a subjective idea. The images of the gods are never to be objects of worship. We should know that they are merely emblems or reminders of various invisible powers and principalities.[8]

Exodus 20:3-6

"Thou shalt have no other gods before me. (v4) Thou shalt not make unto thee any graven image, or any likeness of any thing that is in heaven above, or that is in the earth beneath,

or that is in the water under the Earth:

(v5) Thou shalt not bow down thyself to them, nor serve them: for I the LORD thy God am a jealous God, visiting the iniquity of the fathers upon the children unto the third and fourth generation of them that hate me; (v6) And showing mercy unto thousands of them that love me, and keep my commandments."

CHAPTER 7

THE FASCINATING 777 LINK

Does This Face Represent the Face of God?

777

It was several years ago that the Los Angeles Times, the Pasadena Star News and NBC Television News (among others) first reported on my ocean floor maps discovery. During interviews, one question kept coming up. It is the same question we've been discussing all along: *"Does the face on the Pacific Ocean Floor represent the face of God?"*

Is this image of a weeping man on the Pacific Ocean Floor some holy clue about omnipotence? Can the tears on this face which span almost half of the Earth's surface be the tears of our Creator? Does the face on the Pacific Ocean Floor really represent the face of our heavenly Father God? The story told through the images on our ocean's floor is one of "Good versus Evil." The principal characters are clearly identified with matching Biblical themes.

There is more to this discovery than just characters and story. The theme has the impact of apocalyptic narrative. There are also certain numerical coincidences. One such instance occurs near the western coast of Peru. This is the general area of the mouth-line of the face on the Pacific Ocean Floor. At the mouth on the face on the Pacific Ocean Floor is an overlapping of longitude 77 and latitude 7 (777).

At the mouth on the face on the Pacific Ocean Floor is an overlapping of

longitude 77 and latitude 7 (777).

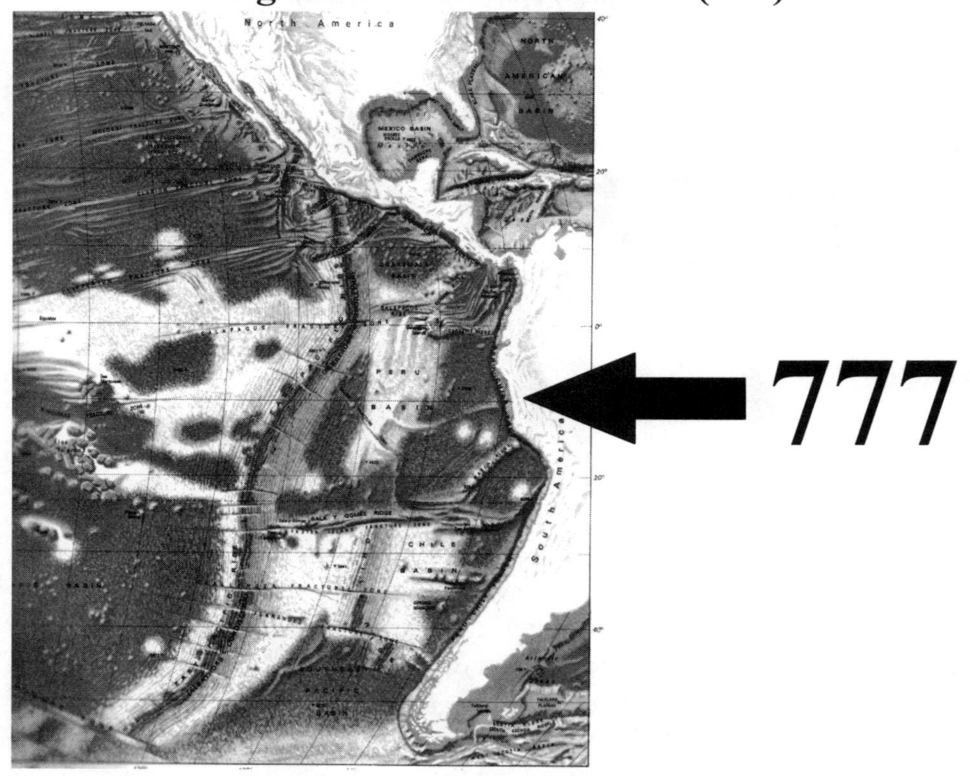

IF GOD'S NAME WERE A NUMBER
IT WOULD BE 777

Some people are curious about the possible significance of this 777 link. What are some of the most logical considerations? There are vast differences between what is commonly called "numerology" and the study of Biblical numerics. Studying the pattern of numbers in the Bible is an old and mostly respected study. It is as old as that of Hebrew kabbalistic thought. At the mouth of the face on the Pacific Ocean Floor is a (777) clue. Also, it is an interesting coincidence that in the Bible, the number 7 is the number of God. Whenever we see three 7's together in scripture, it shows total involvement by God. If God's name were a number, it would be 777.

This is a cropped enlargement of the South American coastal area from the previous page. It highlights the mouth section of a topographic map of the Pacific Ocean Floor. In this book, this map is reprinted in full and in sections with permission by Rand McNally & Company.©

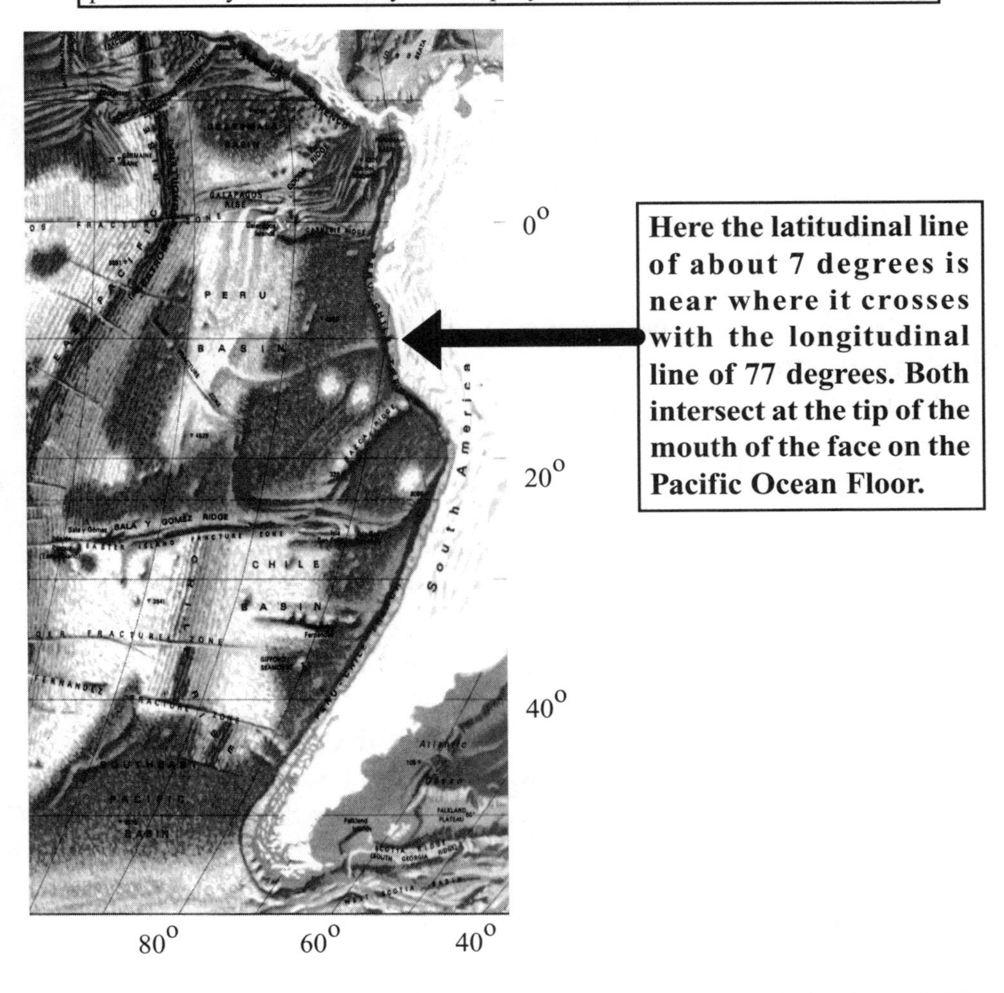

Since a flat map must attempt to represent a spherical earth, the longitudinal lines on a flat map are curved. For example, the 80 degree mark begins at the tip of South America. As the longitudinal line goes north, it passes about 3 degrees from land at what I call the "lip-line," thus the general accuracy of 77 degrees longitude.

Could this curious coincidence have a higher meaning? Could it be a deliberate clue from God, helping us to better identify this riddle? The Creator of heaven and earth is very deliberate in his nature. Numbers and mathematical balance are very important to the mastermind of all spirit and matter. Pythagoras said long ago that "All is number," but God said it first in actions and words. Jesus Christ taught that every word and number in the Bible has permanence. We find this in:

Matthew 5:18
"For verily I say unto you, Till heaven and earth pass, one jot or one tittle shall in no wise pass from the law, till all be fulfilled."

The evidence suggests that Biblical scripture presents a careful and deliberate pattern, using both words and also numbers. There is a mathematical configuration relating to all matter and things, including every human being. God once told Abraham that all of his "seed," his descendants, would be numbered:

Genesis 13:16
"And I will make thy seed as the dust of the earth: so that if a man can number the dust of the earth, then shall thy seed also be numbered."

Identifying patterns of numbers in the Bible has always been a part of good scholarship. Ancient Hebrew writers of the Kabbalah took much of their philosophical and scientific marching orders from the even more ancient Egyptian teachings.

A preponderance of evidence in scripture and nature points to 777 as the most likely number of God. Everywhere in nature the substantiation is self-evident. In scripture we can conclude this because of the overwhelming use of the number 7 identifying God, his actions and his will. The three 7's together further show this because of the Trinitarian nature of God. God in his full essence is triune. He in one nature personifies these three: Father, Son and Holy Spirit.

THE NUMBER 7

No one can be certain of the reason for the 777 link. I personally feel that it is only a coincidence. But what an interesting coincidence and a provocative reason for a more in depth study. But the importance of the 777 link should be scrutinized with care. It might be helpful if we look for the impact of this number in our world. The number 7 is mathematically significant for many reasons. We can point to:

7 days in a week
7 pure notes on the musical scale
7 continents on the Earth
7 basic colors of the spectrum
7 orifices in the human skull

There were the 7 characteristics of wisdom, 7 wonders of the ancient world and Genesis explains that after creating the earth, God rested on the 7th day. Other great religious writings and beliefs seem to agree. Most of them point to the number 7 as the number of divinity. They comprise the characteristics of fulfillment, perfection and completion. For example in the East, it was the Indian yogis that developed the Science of 7 Cultures: health, energy, ethics, will, heart, intuitive skills, and spirit.

The Bhagavad Gita, the Upanishads, the Book of Zoaraster, the Tao, the book of Buddha, the Koran, and the Holy Bible each refer to the number 7 as associated with divinity. But nowhere is the number 7 given such divine importance as in the Bible.[1]

In the book of Revelation, the number 7 occurs more frequently than any other number. In this book alone, we see that there are 7 churches; 7 spirits; 7 lampstands; 7 stars; 7 seals on the scroll; 7 horns and 7 eyes of the lamb; 7 angels and 7 trumpets; 7 thunders; 7 heads of the dragon; 7 heads of the beast; 7 golden bulls; and 7 kings.

Throughout the scriptures, 7 is there to show God's completion. The glory of God's spiritual perfection is evidenced there through his use of the number 7. In the Hebrew language, seven is *shevh*. It is from the root "savah." It means to be "full or satisfied, having enough." Therefore, the meaning of the word "seven" is a complement of this root. For it was on the seventh day that God rested from his work of creation. It was full, complete and perfect.

Nothing could be added to or taken from holy scripture without marring it. It is from the root word *Shavath* (seventh day) to cease, desist and rest, that we arrive at the word Sabbath.[2] The 777 intersection at the mouth of the face on the Pacific Ocean Floor is also symbolically important. Several scriptures provide clues that provoke thoughtful contemplation. Clues are even seen in the life-span of great men and the calculations of events and things.

NOAH AND THE 777 LINK

There is one person in the Bible who is most identified with oceans of water. Noah was saved with his family from the great flood and became the new father of all humankind. He and his family, with their salvaged animals, began the repopulation of the earth. But who was the father of Noah? We might also ask why this is so important. One way that God might hint to the importance of Noah is through a clue about his earthly father. Like Jesus and Enoch, Noah walked with God:

Genesis 6:9
" ...Noah was a just man and perfect in
his generations, and Noah
walked with God."

What about Noah's earthly father? Historians do not really know all that much about him. The Bible tells us that his name was Lamech. After this brief introduction, it tells how long he lived. Here is where we find God's intriguing (777) link to Noah through his father.

Genesis 5:31
"And all the days of Lamech were seven
hundred seventy and seven years;
(777) and he died."

No other person in the Bible lived exactly 777 years. After the flood, every human being in this world died except for Noah and his family. Noah became the new father of all humanity. The lineage of Noah continues with his three sons whose names were Shem, Ham and Japheth.

The two names Shem and Japheth, who received their father's blessing, occur together 7 times; six of these are about Ham whose posterity was cursed. According to those who take numbers in the Bible seriously, none of this is accidental because the words of God are pure.

Psalms 12:6
"The words of the Lord are pure words; as silver tried in a furnace on the earth refined 7 times."

BIBLICAL NUMERICS

During the 1880's, Ivan Panin, a mathematician and former agnostic, began his studies of Bible numerics. Knowing Hebrew, Aramaic and Greek, he began reading the scriptures in their original languages. Greek and Hebrew are unique in that they do not have a number system.They do not use special symbols for their numbers, like our Arabic numerals 1, 2, 3, etc. Instead, letters of the Greek and Hebrew alphabets serve a double purpose and are used to represent numbers.

Aware of the numerical values of the Greek and Hebrew alphabets, Panin began to experiment by replacing the letters in scripture with their corresponding numbers in scripture. By the first day he became utterly amazed. To Ivan Panin, the verses he studied bore unmistakable evidence of an elaborate mathematical pattern. It was far beyond random chance or human ability to construct.

Panin proved to himself that the Bible, in its original language, is a skillfully designed product of a mathematical mastermind. This mastermind, God, made the Bible far beyond the possibility of deliberate human structuring. To prove his point, he supplied the Nobel Research Foundation with more than 40,000 pages of his studies. This was accompanied by his statement that this was his evidence that the Bible is the word of God.

Their reply was revealing and served as a lifelong tool for Panin. It read, "As far as our investigation has proceeded, we find the evidence overwhelmingly in favor of such a statement." Panin then issued a challenge through the newspapers of the world to offer a "natural explanation" or refute the facts. Not a single person was able to do so.[3]

Ivan Panin found that patterns of prime numbers such as 11, 13, 17 and 23 but especially 7, were found in great clusters. He found that the number of words in a vocabulary divides by 7. Also, the number of letters in a vocabulary divides by 7. Additionally, the words that occurred more than once as well as those that only appear once divide by 7.[4]

In 1894, E.W. Bullinger wrote his classic manuscript "NUMBER IN SCRIPTURE."[5] An Anglican clergyman, he was a man of intense spirituality and ardent Biblical scholarship. His effort was to explore the entire field of numbers used in the Bible. His goal was to present his conclusions, *in accord with facts, without needless spiritualizing.* Bullinger's writings were considered free of the excesses that so often characterize works of this nature. Today, some of these excesses are debated concerning a Bible code which foretells events. The evidence concerning such a code is, to me, incomplete and open to much scrutiny.

77 NAMES TO JESUS

Bullinger pointed out many details about the number 7. He believed that the Bible substantiates its Godliness. It became to him the number most worthy of investigation. Not unlike Pythagoras, Bullinger found that the number 7 held properties that were unparalleled by any other number. Bullinger found this to occur in a spiritual way that helped humanity better understand God. Another clue he found is in one of the gospels of the New Testament. There are in the genealogy of Luke, in Chapter 3, exactly 77 names. The first name is God. The 77th and last mentioned is Jesus.

7 WORDS TO CREATION

The ancient Hebrews believed that all things in nature have a pattern. They knew that every pattern could be divided mathematically. Just as the Egyptians used grids in their art, so did the Hebrews in their scripture. Their first statements in the Bible concerning the original creation, consists of exactly 7 Hebrew words. In English they are:

Genesis 1:1
"In the beginning God created the
heaven and the earth."

THE 7 BLESSINGS OF ABRAHAM

It was with great faith that the great patriarch Abraham made his sacred covenant with God. After testing Abraham in many ways, the Lord judged the worthiness of his heart. Beginning with Genesis 22:2, Abraham receives God's "Seven-Fold Blessing":

1. I will make of thee a great nation.
2. And I will bless thee.
3. And make thy name great.
4. And thou shalt be a blessing.
5. And I will bless them that bless thee.
6. And curse him that curseth thee.
7. And in thee shall all families of the earth be blessed.

THE 7 BLESSINGS TO ISRAEL

Also, as part of his covenant with Abraham, the scriptures show God promising a seven-fold blessing to the coming nation of Israel. Each of these blessings begin with the words, "I will."[6]

1. I will bring you out from Egypt.
2. I will rid you of their bondage.
3. I will redeem you.
4. I will take you to Me for a people.
5. I will be to you a God.
6. I will bring you unto the land.
7. I will give it to you.

7 WONDERS OF THE ANCIENT WORLD

1. Artemision at Ephesus
2. The Colossus of Rhodes
3. The Hanging Gardens of Babylon
4. Olympian Zeus
5. The Mausoleum at Halicarnassus
6. The Tower of Pharos
7. The Pyramids of Egypt

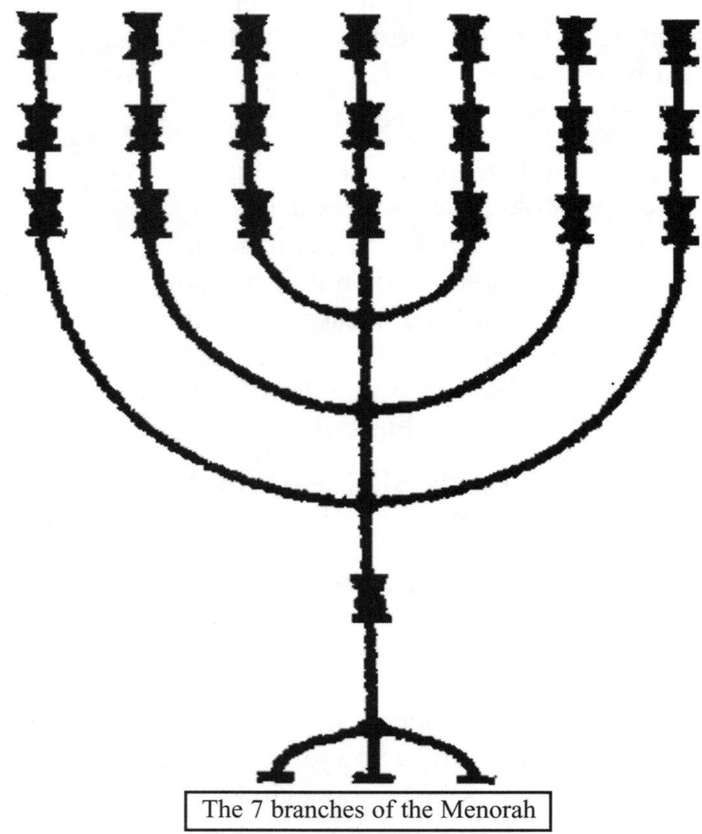

The 7 branches of the Menorah

THE GOLDEN CANDLESTICK

The first obvious teaching diagram of the Hebrew esoteric tradition is the Menorah. This was the candlestick specified by God to Moses on Mount Sinai. Its design and shape was outlined by the Creator Himself. Even how the candles are lit is an established ritual. This symbol has never left its important place in Jewish tradition. At the center of the Menorah is its axis of true grace. The right and left arms are representative of mercy and severity. The Golden Candlestick had 7 branches including the central stem. It represented the light of God and the 7 branches signified that the source of that light was divine.[7]

THE 7 ANGELIC VISITS

Whenever there was an important event surrounding the life of Jesus Christ, God saw that angels were involved. These important servants of the Lord were there whenever Jesus needed them most. But even those visitations were limited to a certain number by the Lord. During the life of Jesus, the Holy Bible teaches that angels appeared here on earth exactly 7 times: [8]

1. To the shepherds (Lk. 2:9).
2. To Joseph (Matt. 2:13).
3. To Joseph (Matt. 2:19).
4. After the Temptation (Matt.4:11).
5. In Gethsemane (Lk. 22:43).
6. At the Resurrection (Matt. 28:2).
7. At the Ascension (Acts 1:10).

THE 7 STATEMENTS FROM THE CROSS

Among the most memorized and quoted phrases spoken by Jesus are those he spoke during the crucifixion. Each phrase was heard in the presence of witnesses. But even these were limited by the number 7 which as we have seen is a number linked to God's completion. There were 7 different statements made by Jesus Christ from the cross, which we know as the "7 last words."

1. "Father, forgive them, for they know not what they do." Lk. 28:34.
2. "Verily, I say unto thee today: Thou shalt be with Me in Paradise." Lk. 28:43.
3. "My God! My God! why hast thou forsaken Me." Matt. 27:46.
4. "Woman, behold thy son...! Behold thy mother!" Jn. 19:26-27
5. "I thirst," Jn. 19:28
6. "Father, into thy hand I commend My spirit." Lk. 28:46.
7. "It is finished." Jn. 29:30.

Of course, there are many other 7's in the Bible that relate directly to God and his completion of things. Actually, there are so many that it would be very difficult to count all of them. This is because a good percentage of these occurrences do not mention the word "seven" specifically. For example:

THE 7 PETITIONS OF THE LORD'S PRAYER[9]

1. Hallowed be thy name.
2. Thy Kingdom come.
3. Thy will be done on earth as it is in heaven.
4. Give us this day our daily bread.
5. Forgive us our trespasses.
6. Lead us not into temptation.
7. Deliver us from evil.

THE 7 VIRTUES[10]

1. Faith
2. Hope
3. Charity
4. Prudence
5. Justice
6. Temperance
7. Fortitude

THE 7 DEADLY SINS[11]

1. Pride
2. Wrath
3. Envy
4. Lust
5. Gluttony
6. Avarice
7. Sloth

There are many other 7's that point to God's involvement. One source is "Strong's Exhaustive Concordance of the Bible." This outstanding reference tool lists the number 7 as being used more than 700 times throughout the Bible.[12]

7 AND THE WORLD'S MAJOR RELIGIONS

Among the world's great religions, ancient and modern, the number 7 is very important. Always present in each of these faiths is the idea of divine completion.

1. The Roman Catholic Church has 7 sacraments Baptism, Confirmation, the Eucharist, Penance, Orders, Matrimony and Extreme Unction.[13]
2. The Babylonian religion showed the Tree of Life as having 7 branches and 7 leaves. The same idea is present in the Jewish candelabra with its 7 branches.[14]
3. Buddha rendered his eight-fold path to enlightenment, but this pattern was based on 7 as a number of completion and 8 then became the number of new arrivals sanctioned by God.

The familiar phrase, "7th Heaven" comes from the Koran as Islam believes in 7 separate Paradises in ascending order of bliss.[15] This belief can also be found throughout Christian history. Some ancient Tibetan traditions have it that man's earliest spiritual teachers were 7 kings who descended from 7 stars.

Hinduism teaches of man existing on 7 planes. They are: Sensation, emotion, reflective intelligence, intuition, spirituality, will and intimations of the divine. Ancient Japanese religion writes of the 7 great Gods of luck. The Mayans believed that their civilization emerged from 7 caves.

There is a term used in the Swahili language called "Qwall." It is used to describe a tradition which celebrates the pursuit of their 7 great virtues. Dante writes that purgatory has 7 levels. To Dante, each level in purgatory has its own special designation. Each level is given its own unique value and individual importance from God. A person's eventual, eternal destination depends on the amount of good works they carry out during their lifetime. Even our passing thoughts are given the weight of most physical deeds. On Judgment Day there will be an accounting for every thought and deed. According to Dante, the final tally will relate directly to the destination of each person's eternal abode. Evil sinners are prisoners in the lower levels which are known as Hades and Hell.[16]

SCIENTIFIC 7'S

Pythagoras, the father of mathematics to some, said this about the number 7: "When examining the numbers 1 through 18, the number 7 is the only number that has certain properties that all the others do not." Pythagoras looked to 7 as central to the balance of nature. Is this why we see 7's in science and nature that also fit with God's plan of completion?[17]

ARE THESE PATTERNS ONLY A CONINCIDENCE?

1. There are 7 natural sciences (grammar, logic, rhetoric, arithmetic, music, geometry and astronomy).
2. There are 7 continents: Africa, Antarctica, Asia, Australia, Europe, North America, South America.
3. There are 7 orifices in the human head.[18]
4. There are 7 major notes in the musical scale.[19]
5. There are 7 days in the week.
6. There are 7 basic colors in the spectrum.
7. Geographers refer to the 7 seas: North Atlantic, South Atlantic, North Pacific, South Pacific, Mediterranean, Indian and Arctic.
8. The Human brain has 7 distinct sections with specific functions inherent to each. They are the Mendulla, the Pons, the Cerebellum, the Midbrain, the Interbrain, the Neocortex and the Frontal Lobe.[20]

OTHER INTERESTING 7'S

1. From China and India come teachings of the 7 chakras.
2. Hippocrates said that man has 7 distinct stages of progression in body and spirit.[21]
3. Shakespeare's famed 7 ages of man.[22]
4. Buckminster Fuller, the inventor of the Geodesic Dome, wrote of the "7 unique axes of symmetry."[23]

"When examining the numbers 1 through 18, the number 7 is the only number that has certain properties that the others do not." *Pythagoras*

WHAT ABOUT THE 777 LINK?

The 777 link at the mouth of the face on the Pacific Ocean Floor does not seem arbitrary. Whatever the reason for the 777 link, it is part of God's plan. Maybe later we'll know more, but for now it remains a curiosity.

TWO HUNDRED MILLION YEARS AGO

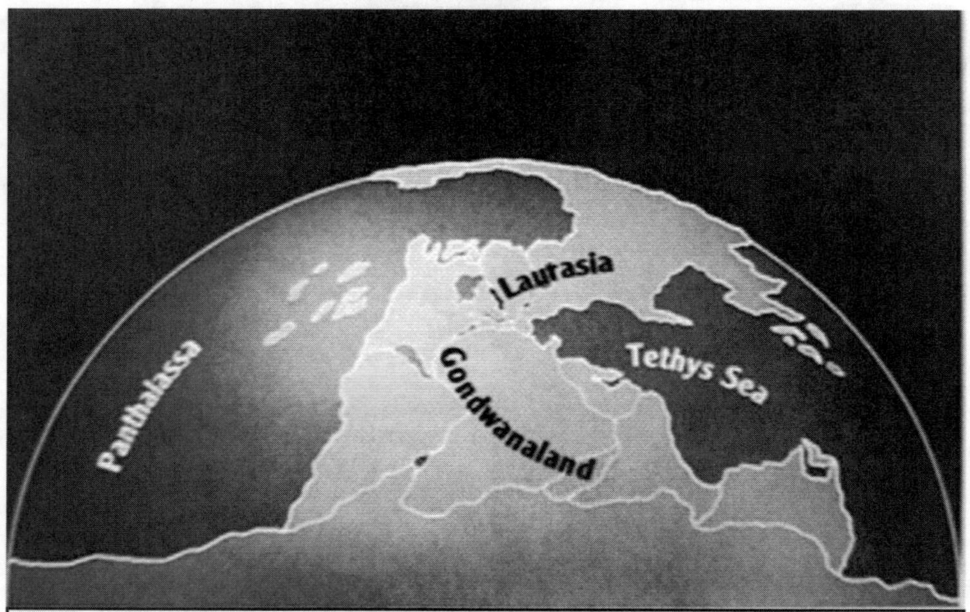

According to the theory of "Continental Drift," this is when the continents were joined into a massive super-continent called Pangaea. Map is courtesy of Celestial Arts.©

LONGITUDE AND LATITUDE

One question I am often asked is this, *"Aren't longitude and latitude very recent in their scientific life, just a few hundred years? How can we compare these lines to the geological shifts in a five and one-half billion year old Earth?"* On the surface it is quite a logical question. How could such recent mathematical calculations as longitude and latitude have any real bearing on geological earth movement? The shifting of ocean floor plates took millions of years. How could Greenwich meantime possibly relate?

LIKE A SLOWLY DEVELOPING PHOTOGRAPH

The one possibility I've considered most is this: The face on the Pacific Ocean Floor is like a slowly developing photograph. This process started towards the beginning of Earth's development and as the years continued, so did the process. Naturally, the image is meant to fully develop more toward the end of that development. Could we be living in that time?

ONE HUNDRED MILLION YEARS AGO

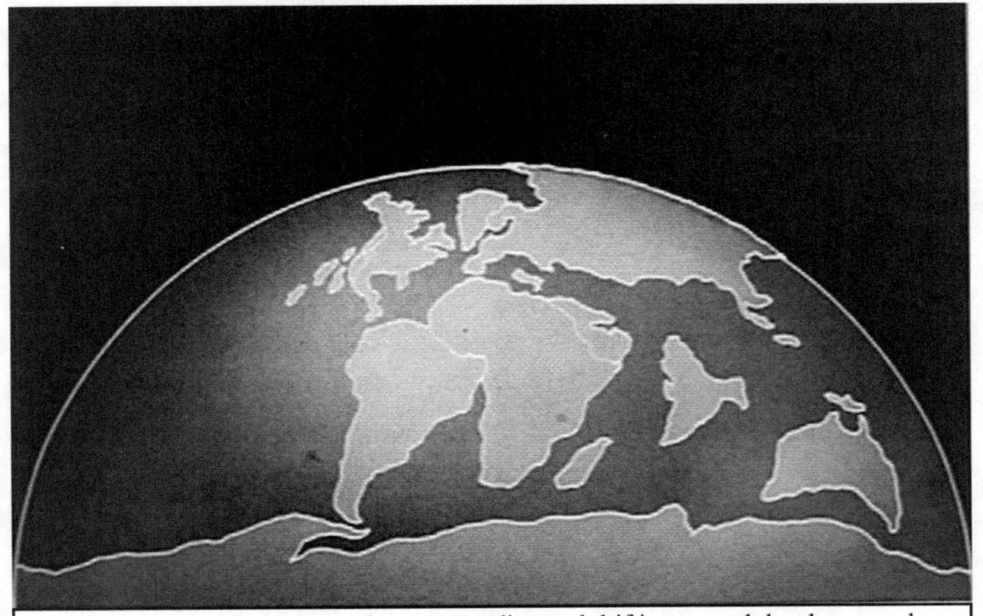

By this time the continents began spreading and drifting toward the shapes and positions that they are today. Map is courtesy of Celestial Arts.©

1675 vs 200 MILLION YEARS

The Royal Observatory at Greenwich Park, England was founded in 1675. Its purpose was to find out the longitude of places for perfecting navigation and astronomy. Before this time, ships at sea had difficulty figuring our their exact location. The Observatory, because its location, would change navigation forever. The official warrant was dated June 22, 1675. The arc distance of longitude measures along a parallel from some starting point. As the earth rotates on an axis that is perpendicular to the planes of the parallel circles, each parallel rotates 360 degrees during each day, or 15 degrees per hour. Because of this, the difference in sun time at one place may be calculated by observing the sun at its zenith. It is difficult to calculate the sun time for two different places simultaneously. *The time at one place must be known at the other location. It is the only way to accurately calculate the longitude.* All meridians are similar and unfortunately there is no natural starting point from which to begin their numbering, as there is with latitude.

With latitude, the natural point of reference for sailors was the Equator. But for longitude there was confusion over which meridian to choose as the prime starting point for navigation purposes. Ocean voyagers were without a common reference.

Then in 1884, a decision was made. The meridian that passed through the Royal Astronomical Observatory at Greenwich was internationally agreed upon as the zero degree or the prime meridian. The length of a degree of longitude varies as the cosine of latitude.[24]

IS THE 777 LINK INCIDENTAL TO GOD'S PLAN?

Since the positioning of longitude and latitude, only began in earnest in about 1884, it is understandable why it seems difficult to see the valid link to 777 at the mouth of the face on the Pacific Ocean Floor. When compared to the age of the earth, any recently calculated man-made point seems incidental to God's plan.

GREENWICH, A RANDOM CHOICE?

Other writers have also wrestled with similar types of difficulties in relating to Greenwich mean-time. Author Bruce L. Cathie experienced a similar stumbling block concerning his research on Harmonics and Earth grid studies. Here is some of what he had to say about this: "All the information available on the reasons, and methods, for choosing the site to mark the position of zero longitude indicated that the choice be a random one, and that the decision of the international body could have placed the zero meridian at any one of an infinite number of positions around the world. I fully believed this to be so until I began to study the mathematical implications of the grid system. I found that many of the geometrical calculations I was experimenting with *were only valid when related to the zero position at Greenwich,* and the appropriate ninety degree intervals around the world."[25]

Cathie goes on to explain, "I was getting what appeared to be valid answers but could not understand the reason why. Many of my critics in certain academic circles were quick to point this out, and argued that the Greenwich meridian was an arbitrary position and so all my work was inaccurate, and that the geometric harmonic values discovered must be in major error."[26]

"Until recently I had no answer to this point. I had to accept the derision and disbelief of those not familiar with the main body of my research. When all the evidence I had amassed over the years was studied as a whole *it indicated very strongly that the zero position was not random,* and that some very positive mathematical process was involved in the choice of this particular line. I fully realized that if my suspicions were correct, the implications were tremendous."[27] (Italics added for emphasis.)

PURELY MATHEMATICAL CONSIDERATIONS?

"If the line was positioned due to purely mathematical considerations, instead of convenience, then it was obvious that the international body concerned was conversant with the associated unified geometric equations just recently discovered in my own research. Who could possibly have been aware of this in the year 1884?"[28]

"The search for the proof I required lasted many years, with the critics snapping at my heels, and deriding my work at every opportunity. I had several verbal battles with so-called experts through the media, and TV, and although I could not hope to win such debates, I believe that I was able to hold my own. Now I believe that I have discovered the necessary proof, which will show the validity of my calculations. The evidence indicates that the Greenwich meridian was meticulously positioned by a group with advanced mathematical knowledge."[29]

Cathie continues to validate his (well thought out) conclusions with mathematical calculations and diagrams. His points (I believe) carry my own rationale one step further. *If the Greenwich meridian has its location because of a group with advanced mathematical knowledge, then the positions of longitude and latitude are not arbitrary.* Could these scientists have had a better understanding of the ancient mysteries than previously believed?

The advanced mathematical calculations Cathie refers to concern the natural harmonics of the earth, treating it as if the earth were a giant natural tuning fork. To me, it helps validate the idea that the 777 link at the mouth of the face on the Pacific Ocean Floor, could be considered deliberate. This may be possible because of natural energies that have always been understood by God and until only recently understood by man. Every aspect of nature we learn about has with it an aspect of patterns.

Author Lucie Lamy links the importance of ancient thought regarding numbers, as they relate to all things, with modern atomic science. She writes, "The Pythagorean axiom, 'All is Number' was long considered an unrealistic proposition, until it was discovered that the properties of any given chemical elementare governed by a specific whole number: The atomic number, corresponding to the number of protons and electrons belonging to its atom, as well as their distribution on the orbits (levels of energy) surrounding the nucleus. Hydrogen for example, has an atomic number of one with only one electron in orbit, while helium has the number two and two electrons. The transmutation of hydrogen into helium is at the origin of the birth of stars and in our small world, provides the solar energy from which we benefit."[30]

Considering the preceding information, it's easier to appreciate why the 777 link at the mouth of the face on the Pacific Ocean Floor does not seem arbitrary. Whatever the reasons for the 777 link or coincidence might be, they are part of God's plan and remain a mystery.

Chapter 8

THE FACE ON MARS
(and other images)
How Do They Compare With
THE FACE ON THE PACIFIC OCEAN FLOOR

If I can accomplish anything with this chapter, it will be to establish my own general skepticism. During presentations and interviews, I am often asked about other kinds of images seen in rock formations, window glass and clouds. Evidently, there are millions of believers in the occurrence of such images. Most of them are sincere and even devout.

The reasons why these people believe what they believe are often based on faith or desire. This is understandable to a point. God and his many manifestations are limitless. But my study does not have the luxury of arbitrary subjectivity. While the color and shading of the map tracings do involve some artistic license, this must be done with integrity. Embellishments must not corrupt the theme of the map tracings.

THE STRENGTH OF ITS SIMPLICITY

Objectivity and respect for science must outweigh any suppositions and religious beliefs we may have. How can we scientifically link God and the Bible to the images on our earth unless science gets full respect? The substantive goal has always been first to show that these images are real. If they are clues from God, then doubters must explain why they are not.

Some folks might wonder how I can respect science and yet end with theology. To me, in their purest form, both are the same. Science objectively carried out, leads to the reality of God. Most endorsements concerning this study are in response to the *strength of its simplicity.* For the most part it is an equal-opportunity discovery. The guidelines are simple, logical and structured. They are easily interpreted regardless of age or intellect.

ONE MAJOR RULE

Find any small child who knows how to trace pictures and tell that child to start tracing these maps according to one rule and their results would be the same. That one rule is this: *They can only trace where the ocean shore meets the continents.* If an expert in any scientific field were to follow this one rule, the results would be the same. If somehow it were possible to train a monkey, or a dog, or even a parrot to trace, the results would be the same.

If a test could be carried out where illiterates who know how to trace could join nuclear physicists, aeronautical engineers and grammar school students, all participants could get the same grade. The reason is simple. *I have never met anyone, who when sticking to this one simple rule, did not come up with similar images.*

GUIDELINE FOR SECONDARY TRACINGS

After the simple border tracing takes place, there is a second phase, which only adds to the clarity of the images. Trace the highest and deepest underwater mountain ranges. When this is done, facial and body characteristics show an enhancement. Included would be the eyes, teeth, bone structure and muscle tone. Following the one main rule above and the guideline for secondary tracing will always result in the same four sets of ocean floor images in this book.

This is not to say that, in the future, other methods might not be incorporated. I'm sure they will. In a way, any guideline for tracing these ocean floor maps is practically moot. The details in these maps are clear. There really is no need for an accompanying border tracing. If the *integrity of these images could not stand alone,* they would not have such extraordinary impact.

Over the years, I have spoken to many people concerning other images that are both terrestrial and extraterrestrial. In that time I have learned that anyone can see a face in the clouds. How and why we see what we see matters. That is why so much is said about perception (page 36) in Chapter 4 of this book. I have continually responded to such examples cautiously. Often people show me stories and pictures they've seen published. Most are not very credible. Still, I feel that it is important to tell about some of these images.

I deliberately do not reprint pictures of the following examples, but most have footnotes. All anyone need do is imagine each likeness (at its best), and he would probably be close. Each may have validity (with additional evidence), but that additional evidence *is still not seen.* The simple fact that these things look like what people say they look like is not enough. Still, they can be amusing and sometimes interesting.

THE FACE OF JFK ON A MOUNTAIN

Whenever I discover or see such images, I always look for relating factors and additional evidence. Take for example on the Hawaiian island of Maui, there is a landmark image of John F. Kennedy's profile on a mountain side. It is located at Black Gorge in Iao Valley. Postcards showing a picture of the image are available to tourists. The belief in this image has grown to such a degree, that they've even set up a special viewing scope there for viewing the image at its best. On any day of the week, tourists and locals study it. They wait in line to look through the scope. They can't take their eyes away from the image of John F. Kennedy's face.

This example is exactly what the face on the Pacific Ocean Floor is not. The main appeal of this tourist attraction on Maui is that it really does look like JFK. But that is it! There is no additional evidence that this is true. There are no comparisons given or any additional evidence shown. It could be the face of any man who looks a lot like JFK or it could be only a silly coincidence.

Of course, it is most likely just coincidence. People have fun with it and most folks don't look at such things very seriously. They just look and say, "isn't that interesting?" But suppose for a moment that the evidence did not stop there. What if there were more to the picture. Suppose, not only did we see the image of the head of JFK, but also a bullet coming toward it. And on the mountain next to it, what if we could see the image of a guy who looked just like an assassin, pointing a gun at him? Then, the likelihood of this image depicting Kennedy would grow.

After that, what if we could come in close on the head wounds and see that they matched the actual pictures of a wounded JFK just after his assassination? Can you imagine the reaction? The images on our oceans floor are many times more real than any such pictures.

JESUS IN THE CLOUDS AND SNOW

One small booklet ambitiously includes several stories concerning sightings of the Virgin Mary. It also displays on its back cover a photograph of what is described as the face of Jesus Christ. It is seen naturally emanating from a snowy hillside. Inside is another photograph. It shows the front page of a Corvalis, Oregon, newspaper that is dated December 24, 1964. This large photograph nearly fills the upper fold. Above it is a boldly printed headline that reads: "STRANGE CLOUD FORMATION." The photograph of the cloud formation shows what is supposed to be the image of Jesus Christ. It shows the front view of the upper part of his body. His arms are extended and clearly visible. This image in the clouds also includes such details as his mustache, hair, eyes and fingers.[1]

THE ALCOHOLIC'S NIGHTMARE

Professor Wilson Bryan Key's book, "The Clambake Orgy," reveals horrible images, subliminally painted in the graphics of ice cubes in liquor print advertisements. He shows many examples of drunks, skulls, monsters, ghosts, nude women and demons. He also shows a picture of a plate of clams from a restaurant chain's menu. The author writes that it is actually several nude men and women involved in a sex orgy.[2]

JESUS AT MOUNT ST. HELENS

One national Protestant Christian television ministry provides 8x10 color copies of a photo they say is the face of Jesus. The face of a bearded man comes from a large cloud of ash emitting from the eruption of Mount St. Helens. Much mention is made by the ministry that this photographic phenomenon took place on May 18, 1980. The Pastor of this church also teaches that this was the anniversary of Pentecost day.[3]

MARY ON A WINDOW PANE

Thousands of devotees, mostly Catholic, showed up outside a home in Bakersfield, California, to see an image of the Virgin Mary on a window pane. The story with pictures ran on most west coast television stations. Dozens of people gave testimony about their beliefs in this "miracle."[4]

JESUS IN A SANDSTORM

One unverifiable story claims that the Catholic Church officially researched an aberration which occurred in Somalia, Africa, in 1993.

Hundreds of people witnessed what many said was a five-hundred foot image of Jesus hovering in the clouds of a sandstorm. Someone took a photo of this giant face and presented it to the Catholic Church. So many people described the image as the face of Christ that the Catholic Church has become officially involved.

A Vatican source said, "We are taking this vision of Christ very seriously. The Pope personally ordered a full investigation less than an hour after we heard the initial reports." He now believes that the image was "God's way of telling mankind that we should be helping each other instead of killing and maiming."[5]

THE GREAT STONE FACES OF UTAH

In the Red Rock country of southeastern Utah, there are many rock formations that look like faces. One magazine published several different photographs of these images. Each photograph had with it names like, "Stone Face," "Sharp Young Man," and "The Moon Watcher." Others look like deer, penguins and sheep.[6]

QUEEN VICTORIA'S MOUNTAIN

Several years ago famed atheist, Madalyn Murray O'Hare, invited me to her annual Winter Solstice Dinner. After dinner she and several of her disciples met privately with me and an associate. I shared with them the face on the Pacific Ocean Floor. Expressing fascination, she told me of a unique mountain in England. It evidently looks like Queen Victoria. The people there call it "Victoria's Mountain." This type of phenomenon is not unusual. I have many examples of similar stories like this on file.

SKULL HILL AND THE HARP SHAPED LAKE

The famous hill in the Bible relating to the crucifixion of Jesus is located at Golgotha. It was called "the place of the skull."[7] Modern pilgrims have named it "Skull Hill." When I was in Jerusalem, I learned the reason why it has this name. It is because Skull Hill has a shape similar to a human skull. It is the same reason why the Sea of Galilee is called, "The Harp Shaped Lake." Its shape is like a harp. It is also how the Oklahoma-Texas Panhandle got its name. On a map, it looks like a pan with a handle.

Such thinking also accounts for how the "Big Dipper" got its name. It looks like a giant soup dipper with a handle. There are thousands of such examples all over the world and even into the far reaches of outer space. Even the signs of the Zodiac have their names from what they looked like in the night sky to ancient star gazers. To them, the constellation of Leo looked like a lion, Libra looked like scales, and Cancer like a crab. Even modern astronomers can point out image similarities in most of the constellations.

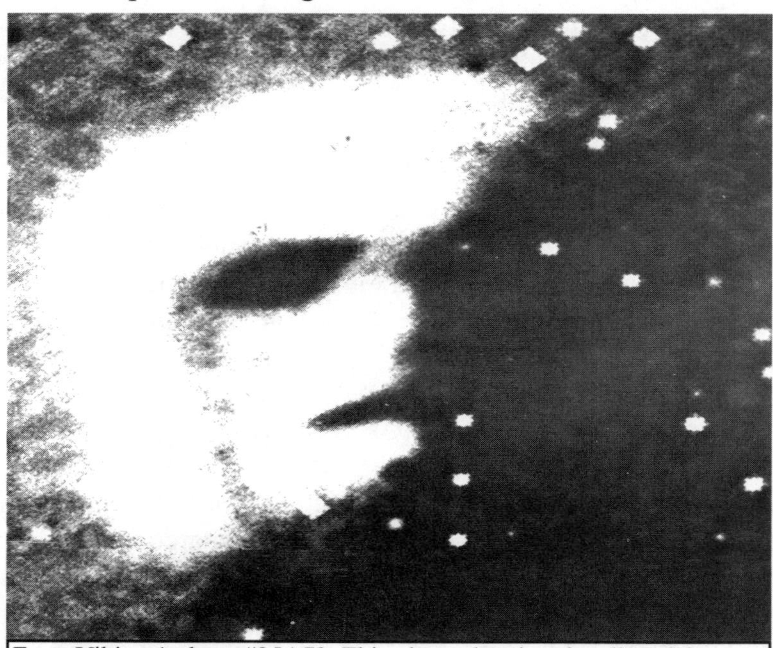

From Viking 1 photo #35A72. This photo showing the alleged face on Mars is shown courtesy of the National Space Science Data Center.

THE FACE ON MARS

Another image that has been getting lots of attention in print and even on the Internet is the so-called "Face on Mars." Often, people ask me about this face and I have done some research on it. I've spoken to or interviewed many key researchers involved and I have carefully studied the NASA and JPL data. Although, much of this research is still ongoing, I believe more proof is necessary. *Until we can get clearer pictures or more data, the face on Mars can only remain a curiosity.* Until there is another Mars mission, resulting in clearer pictures of the area, I remain a doubter.

Still, since it has created so much interest, something more should be said. *But my interest in the face on Mars is only as it relates to the face on the Pacific Ocean Floor.* The stories, research and interviews relating to this one-square mile area on Mars continue to appear. For now, I will keep updating my files.

One story tells of a protester picketing outside the Pasadena based Planetary Society. This is a grassroots organization of scientists searching for evidence of extraterrestrial intelligence. This protester told the society's director that he is upset there has been no formal investigation of the Martian mountain that resembles a face. Such protests are not taken very seriously. NASA likens this image to the face of the "Man in the Moon." One Planetary society official pointed out that the image looks a "lot like Elvis."[8]

I first began researching the face on Mars in 1981. That is about when newspaper and magazine articles started appearing. One article told about two Goddard Space Lab scientists that are pioneers in this study. Their names are Vincent Di Pietro and Gregory Molenaar. Shortly after the stories began appearing, I spoke with Vincent Di Pietro on the phone. He told me he felt the scientific community should do more research concerning the face on Mars. He was kind enough to send me their manual entitled, "Unusual Mars Surface Features."

I sent him press clippings and general information concerning my discovery of the face on the Pacific Ocean Floor. Later when we talked again, he told me more of his frustration with the scientific community and how hard it can be to get their attention. He explained that he was not asking for a leap of faith from anyone, just more careful study. His advice helped me lecture more confidently concerning my own research.

Their computer enhancements concerning the face on Mars were cutting edge for 1982. Di Pietro and Molenaar were meticulous in digitizing the data. They configured the decimal value of each bit, reducing computer words of one and zero. The result was an understandable gray scale. With that, they were able to stretch the contrast of the face on Mars, resulting in better picture clarity and resolution. Using a special split method, the results were startling. The features near the mouth and nose on the face became more easily visible.[9]

Some of their followers consider the research of Di Pietro and Molenaar similar in ways to that of Galileo and the difficulties he had with skeptics. Galileo proclaimed that there were craters on the Moon and proved the point with his telescope. His evidence was the detail of the craters seen through his telescope. Although they were unmistakably craters, the technology of lenses was still young. Many people did not know how to interpret what they were seeing. So the peers of Galileo, and especially the Church, only scoffed at his discovery.

Di Pietro and Molenaar proclaimed that the face on Mars is relevant because of its detail. They proved the importance of that detail by using innovative computer graphic enhancement technology. Even now, this method is young. History has exonerated Galileo. Maybe in time, it will also exonerate Di Pietro and Molenaar. Other researchers have continued and sometimes paralleled the work of these two men. Most notable would be Richard Hoagland. His book, "The Monuments of Mars: A City on the Edge of Forever," shows respect for the work of Di Pietro and Molenaar. It also contrubutes additional details. Hoagland's complete study of the face and the "city" surrounding it, is comprehensive.[10]

Hoagland proclaims that his research offers evidence of an ancient Martian civilization. The author lays out a grid of the entire plane next to the face on Mars. He displays so much detail that we soon realize this face is only a small part of a very complex terrain. The computerized three-dimensional rotation of the face on Mars is especially impressive.[11]

Hoagland's study of Mars images developed into "The Mars Project," which has the membership of several Berkeley professors. The historic exhibition was presented and called "First Contact the Search." The three month exhibition by Scottsdale's Municipal Center for the Arts ended in February 1987. It brought together NASA's "SETI" project and the "Mars Project." The result was a variety of perspectives on the search for extraterrestrial life.[12]

SCIENTIFIC SKEPTICISM

What do planetary professionals think of this effort? "Not much" says Harold Masursky, a member of the Viking team. "You can find 'faces' in any number of terrestrial mountain ranges, and I can show you better formed 'pyramids' in aerial photographs of Arizona."[13] Carl Sagan considered the "scientific" scenario by Hoagland and company too flawed. He asked them to say, if they list his name in their paper, that his support is *less than complete.* Specifically, Sagan said the possibility of an ancient civilization is the "least likely explanation" of the Martian features.[14]

Words of skepticism concerning the face on Mars are greatest by the very keepers of these images. One statement comes from a scientist from NASA's Jet Propulsion Laboratory, the custodian of the Viking images. He says "the twists and turns of the Martian surface are like a Rorschach test. This face can be read in almost any way, even as Kermit the Frog."[15]

Still there are others that find fascination with the topography of the Martian plane as it relates to possible present or past life on Mars. One speaker on the subject is Educational Psychology and Technology Consultant, Edith O'Donnell. She presented an extensive lecture and slide show in February, 1991, at the Pasadena Convention Center. O'Donnell's topic was "The Face on Mars and the Extraterrestrial Connection."

A NASA PLOT?

After her lecture, I had an opportunity to conduct an on-camera interview with Mrs. O'Donnell. She said she feels that certain unknown people at NASA, deliberately want to avoid lending credence to the images on Mars. Although she covered much of the same ground as Hoagland, Di Pietro and Molenaar and others, she added her own slant.

I asked Mrs. O'Donnell how important she felt the phenomena of these images are. Her position is that this topography may be the first verifiable proof of past or present life on Mars. She further urges that solid proof can only result from a physical landing on this Martian Plane. "At the very minimum, a low altitude fly-by would resolve this question," she said.[16]

In late August, 1993, the nearly $1-billion Mars Observer Project, which was to resolve many unanswered questions, failed miserably. This was to be the first U.S. mission to the solar system's mysterious fourth planet in seventeen years. The Mars Observer, at the crucial moment, just outside the planet's atmosphere, ignored NASA'S commands and evidently missed its appointment with the Red Planet. NASA flight controllers didn't know whether the planetary probe had safely reached a planned Martian orbit on its own or had disappeared in the void between planets. NASA staff had nothing except a faith in their own human engineering skill to sustain the belief that the spacecraft still existed at all.

The failed mission of the Mars Observer only added to the mistrust in NASA. Those who believe that NASA or our government has something to hide are more upset than ever. Even with lots of news relating to the 1997 Mars landing, very little is said by NASA concerning the face on Mars.

So far, I have not had to worry about any plot or scheme to hide or discredit the face on the Pacific Ocean Floor. Maybe it is because our Earth's topography is so much closer to home. Maybe it is because the details on these topographic maps are too easy to verify. When people ask me to compare the face on Mars, as it relates to the face on the Pacific Ocean Floor, I am reminded of a past interview on this subject.

HOW DO THEY COMPARE?

Several years ago, I produced a one-hour documentary about the images on our ocean's floor. It was my main project for the year when I was a senior at Cal State at Los Angeles. I called it, "Scientific Proof of the Deliberate Supernatural." For the documentary, several scientist and theologians were interviewed. Also we interviewed experts in oceanography, geography and mathematics. It was well received and became a submission by the University for national awards recognition.

We interviewed attorney and Baptist Minister Dr. John Stewart. At the time he was a national radio talk show host with the Christian Research Institute in Southern California and a professor at the noted Simon Greenleaf School of Law. He also was among the first to learn of the face on Mars. I wondered what Dr. Stewart thought, when comparing the validity of the face on the Pacific Ocean floor, to that of the face on Mars. The interviewer is Michael Ellington. This is part of the interview relating to this comparison.[17]

Ellington:

A few years ago you had an opportunity to be among the earliest people to witness this special photograph. This photo from NASA is a one square mile area on Mars of a face of some kind. What circumstances involved that event and what were your feelings concerning that scientific discovery of a face on the surface of Mars?

Stewart:

Well, there are two individuals who have spearheaded the interest in this face on Mars, Di Pietro and Molenaar. There is a (one-mile-across image) that the Jet Propulsion Lab calls a 'Monkey Face,' although Di Pietro and Molenaar think it's a human face. I've seen the pictures that show some amplification by computers to suggest that this may very well be some artificial figure. Now, what it is, how it got there, or whom it represents, those are various unanswered questions. It will take another Viking probe of Mars or something similar like that to answer them. But there may be some parallels if we can demonstrate them, but that really is a big 'if.'

But are we able to demonstrate that this face on Mars is artificial, man-made or represents something other than some natural phenomena? So, we could perhaps link (this face on Mars) to what we're talking about today, in terms of the ocean basins. But the connection yet, isn't clear and we need more evidence.

Ellington:

If you were to compare or contrast your feelings when you first learned about the face on the Pacific Ocean Floor and contrast that with the imagery you first saw from NASA of the surface of Mars, were your feelings different at all?

Stewart:

It would be hard to compare precisely but I would say that Mars is more or less a long shot. We need a lot more information, whereas we do have the (ocean floor) maps now. We can do further studies on perhaps statistics about how these forms could have gotten here. If by coincidence and if not, that is when we need to give an explanation. Who put these images here? Why are they here? I would say that the earth's phenomena are certainly much easier for us to discuss, closer to home.

Ellington:

What do you think about the fact that the ocean floor maps on earth look like the images they do and relate as Lloyd Carpenter asserts. Are these images on our ocean's floor examples of the deliberate supernatural?

Stewart:

I'd say we can't rule that out. We need to be cautious. We should apply scientific and historical methods to go through all of the various possibilities, but do it with an open mind. Not necessarily with a religious bias to read into it what we want, but to carefully ask certain questions. Could this be something that is designed to tell us some things about God? We shouldn't dismiss this out of hand. The evidence seems to be such that someone who is cautious and careful won't dismiss this out of hand, but will at least look further into the evidence that we have.

LIFE ON MARS?

On August 7, 1996, a story appeared on the front page of the Los Angeles Times concerning the possibility of life on Mars. Both NASA and Stanford University scientists announced that they had discovered evidence that life may have existed on ancient Mars. They stated that a Martian rock which was found on Antarctica in 1984, contained fossils of single-celled organisms similar to ancient life forms on earth. Estimated at 4.5 billion years old, they said this rock was blasted off the surface of Mars when an asteroid crashed into the planet about 15 million years ago. This was precisely thirteen years to the day (August 7, 1983) that this same newspaper made reference to my discovery of the face on the Pacific Ocean Floor, also on its front page.

As of this writing, ongoing missions include the Mars Global Surveyor which will make a map of the entire planet and is able to resolve structures as small as a car.[18] Once this mapping project is completed, maybe then figures such as the face on Mars will be reconciled. Could this also eventually lead to more answers concerning the mystery of the face on the Pacific Ocean Floor? Only time will tell.

Chapter 9

THE DEVIL IN THE DEEP BLUE SEA
(And the 666 Link)

This gargoyle appearing creature below is seen when combining two Rand McNally© topographic maps. They are the South Atlantic Ocean Floor and the South Polar Ocean Floor. Let us call this possible serpent the Devil, because we can see that he is filled with hate and *attacking the weeping man on the Pacific Ocean Floor* by the throat. His fangs are located at the tip of South America. At his neck is the biggest chunk of ice in the World.

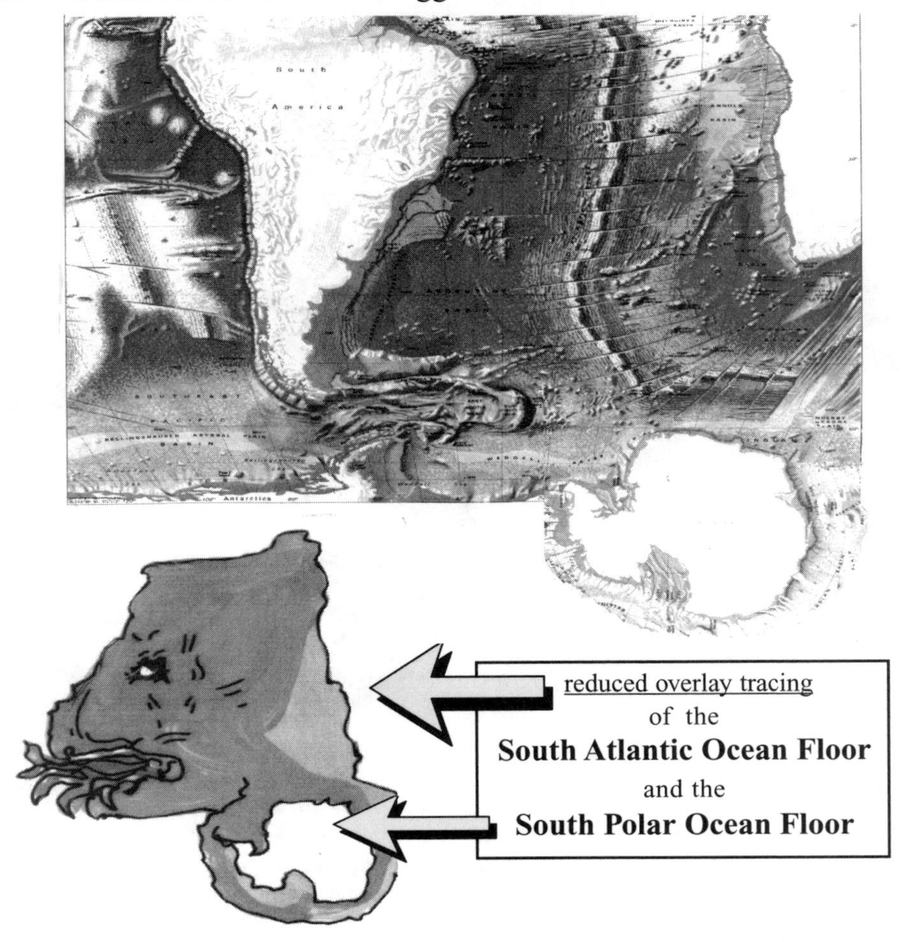

This page shows a conceptualized border tracing of the two topographic maps connected. We call this image:

The Devil in the Deep Blue Sea

The Devil dominating the South Atlantic is either ripping at, or breathing fire into the neck of the weeping man on the Pacific Ocean Floor. At his neck is a sea of ice, the South Polar Ocean. We can see from his expression that he is filled with hate.

This ocean floor map focusing on the

SOUTH ATLANTIC OCEAN FLOOR

Appears more real than any tracing could ever be.

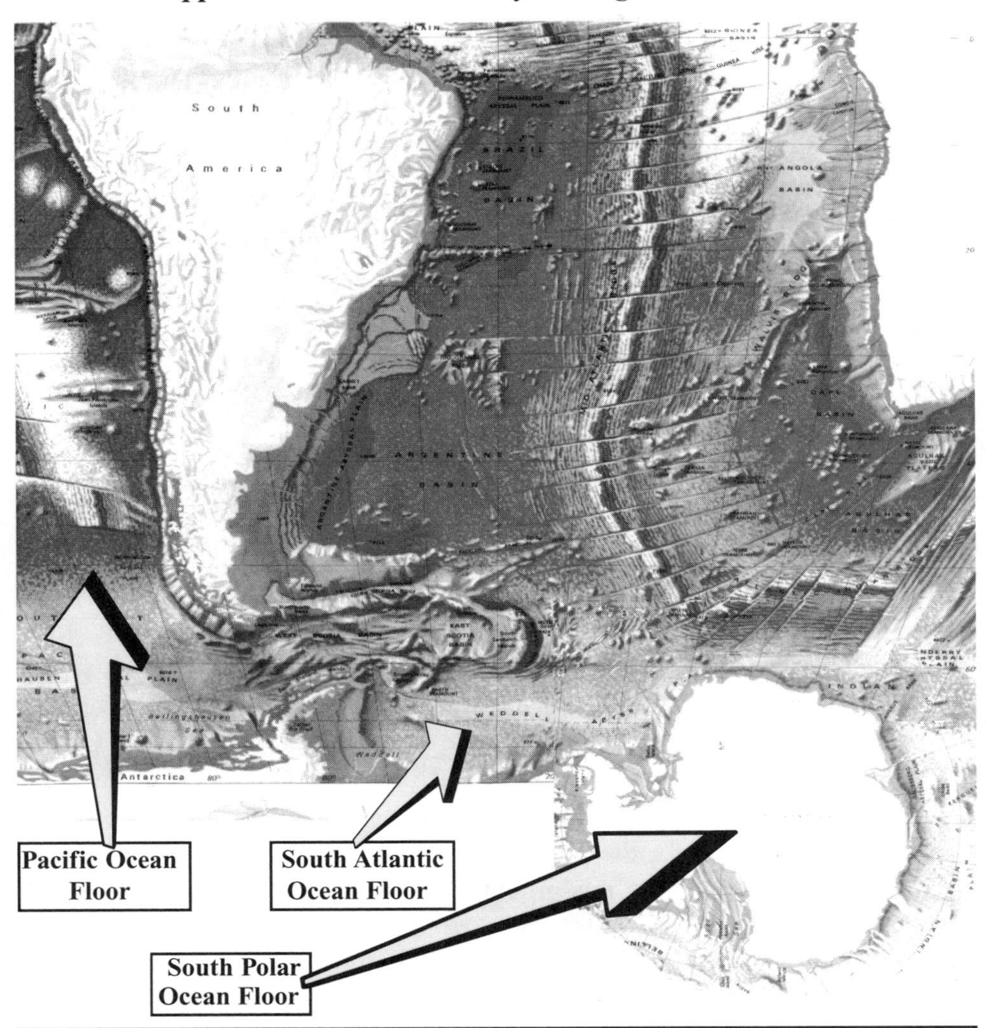

We attached the map of the South Polar Ocean Floor to this map of the South Atlantic Ocean Floor to see the proximity of their positions to each other. Both maps are produced by the Rand McNally Company.© At the left of the west coast of South America we see the mouth, chin and neck of the face on the Pacific Ocean Floor.

THE 666 LINK

This provocative image shows an interaction between the face on the Pacific Ocean Floor and the demon. We see that this devilish character is attacking the weeping man on the Pacific Ocean Floor. His fangs are at his throat. At approximately 66 degrees longitude and 60 degrees latitude (666), he is attacking the neck of the man on the Pacific Ocean Floor. This (666) link is found at the lower tip of South America, in the West Scotia Basin, just south of Cabo de Hornos.

His demonic eye is actually the entire Bromley Plateau Grande also known as the Rio Grande Rise. We see the interaction of this beast with the weeping face on the Pacific Ocean Floor. Considering this (666) link invites speculation. It is a reminder of the beast prophesied when our world comes to an end.

Revelation 13:18

"Here is wisdom. Let him that hath understanding count the number of the beast: for it is the number of a man; and his number is Six hundred threescore and six."

THE ABSOLUTE OPPOSITE

Contrary to the mercy and love shown in the expression of the face on the Pacific Ocean Floor, the demon is grotesque. He is blinded in his rage and vicious in his manner. He is on the attack of all that is good. Contrary to the gift of grace, he offers hate and malice. If the face on the Pacific Ocean Floor is supposed to be a depiction of our heavenly Father God, then this creature is none other than the Devil.

When we connect the topographic map of the South Polar Ocean Floor (at Antarctica) we see this gigantic chunk of ice. Its location is at the throat of the demon. Considering that Dante and others have described the Devil as living on a sea of ice, the comparison does not go unnoticed.

One section of the Dead Sea Scrolls describes Satan breaking out into the open, evidently (in part) from the bottom of the sea. It reports that the "deeps of the Abyss shall groan amid the roar of the heaving mud." It then describes a terrible earthquake and predicts that all of the Oceans in the world will "howl."

The Mouth of the Devil on the South Atlantic Ocean Floor is near

66 degrees longitude and 60 degrees latitude

He is attacking the throat of the weeping man on the Pacific Ocean Floor.

The Father Of Lies

John 8:44

"Ye are of your father the Devil, and the lusts of your father ye will do. He was a murderer from the beginning, and abode not in the truth because there is no truth in him. When he speaketh a lie, he speaketh of his own: for he is a liar, and the father of it."

The Dead Sea Scrolls also make the prediction that when Satan breaks out, everyone traveling on the ocean then will die. Then the voice of God cries out and the planet will "stagger and sway." Then the Angels begin scourging the earth with the worst fire in human history.[1]

THE WRATH OF GOD

The Bible has many verses which agree with the scenario presented in the Dead Sea Scrolls. Future writings will go into this area more in depth. But during this same period, God's wrath will not only be directed at the Devil but also all the ungodly people in the world. Also, everyone will understand why God reveals his wrath. Nothing will be hidden anymore. There will be no excuses.

Romans 1:18-19-20

"For the wrath of God is revealed from heaven against all ungodliness and unrighteousness of men, who hold the truth in unrighteousness; Because that which may be known of God is manifest in them; for God hath shewed it unto them. For the invisible things of him from the creation of the world are clearly seen, being understood by the things that are made, even his eternal power and Godhead; so that they are without excuse...."

The above two tracings dramatize the totally different personalities involved. Notice that we see only the *left side* of the attacking beast. His evil expression tells us all we really need to know about his demeanor. He is absolutely opposite in nature to the weeping, loving man on the Pacific Ocean Floor. He is the epitome of evil incarnate.

WHY IS THE DEVIL SO ANGRY

At one of my lectures, I noticed that a person in the group was especially interested in the map tracing of the Devil on the South Atlantic Ocean Floor. During the question and answer period, he asked me some questions I hear frequently. He said, "How come the Devil on the South Atlantic Ocean Floor looks so angry? What is it that has gotten him to lose his temper and go into such a blind rage? Why is he attacking the throat of the weeping man on the Pacific Ocean Floor?"

I told him that it is probably because Satan hates God. But to me, the real benefit of the image of a Devil on the South Atlantic Ocean Floor is not that his emotions are so easily decipherable. Nor is it because we are curious to know the reasons for his blind rage. The most important thing about this image is that it confirms the possibility of there being a Devil at all.

We've shown how the face on the Pacific Ocean Floor can be evidence that there really is a personal God. I believe also, in the same way, the demonic image on the South Atlantic Ocean Floor can prove to even the most skeptical of researchers that there really is a Devil. I'll give you an excellent example of what I mean.

A few years ago my wife and I were returning from a summer vacation in Mexico. As necessity would have it, we had to spent a few extra hours at the airport terminal near Puerto Vallarta. We were waiting for our flight back to Los Angeles which was delayed. Sitting next to us, in the waiting area, was a young man in his early 20's. He was wearing short sleeves, like most of us, to overcome the tropical heat. I noticed that he had a tattoo about two inches high prominently displayed on his upper right arm. The tattoo was of a little red Devil with a pitchfork. The little Devil wore the expression of a mischievous looking snarl. This cartoon tattoo was not designed to be scary or threatening. It was almost cute.

THE LITTLE RED DEVIL

Since the young man was part of our tour group and no stranger, I felt comfortable asking him about his tattoo. After complimenting the quality of the artwork, I asked him why he chose this image to be put on his arm. He told me that he and several friends all got tattoos at the same time. He said that he thought this little red Devil cartoon was interesting. So that is what he selected rather than pick a tattoo of a heart or an eagle or whatever.

Then I asked him, “Do you believe in the Devil? Do you think that there really is this evil being known as the Devil?” He responded quickly. “No way!” He went on to explain that he had been raised as a Catholic and was taught all about God and Jesus and the Devil but he didn’t believe in “any of that stuff.”

His comment made me curious. If he didn’t believe in the Devil, why would he choose to put a tattoo of the Devil on his arm? He explained that this was exactly the point. He said, “If I ever really thought there was an actual real Devil, I’d never be stupid enough to put his picture on my body. My tattoo is nothing more than an innocent sign of rebellion. This is a purely mythical character.”

I wasn’t surprised at the response he gave me. Most surveys report that lots of people don’t actually believe there is a real Devil or that the Devil wants our souls so he can make us his slaves in Hell. Pastors and Priests everywhere agree that one of the greatest tricks the Devil can play on a person is to convince them that he (the Devil) doesn’t exist. If Satan can get you to believe that he is not real and that there is no Devil, then that makes his job all that much easier.

Since we had plenty of time before boarding, I showed my maps and their accompanying border tracings to him. I also let him read some of the newspaper articles we had with us. After reading and studying the material and asking several routine questions, he responded in amazement. “I first thought you were crazy talking about these images. But when you showed me that big giant face on the Pacific Ocean Floor, I was flabbergasted. Then when I saw the Devil on the South Atlantic Ocean Floor, I realized that there are too many elements for this to just be an accident.” Later, he swore that he would have his tattoo removed as soon as he got back home. “The last thing in the world I want is a permanent picture of this stupid little red Devil tattooed to my arm.” Then he said, “I feel that way, especially now that I think there might really be a Devil!”

No human being created the face on the Pacific Ocean Floor or the Devil on the South Atlantic Ocean Floor.These are sculptures in stone. People instinctively know that these images could only be created by an all powerful God. By their very fantastic nature, such border tracings draw even the most agnostic thinker into a more personal attitude about God.

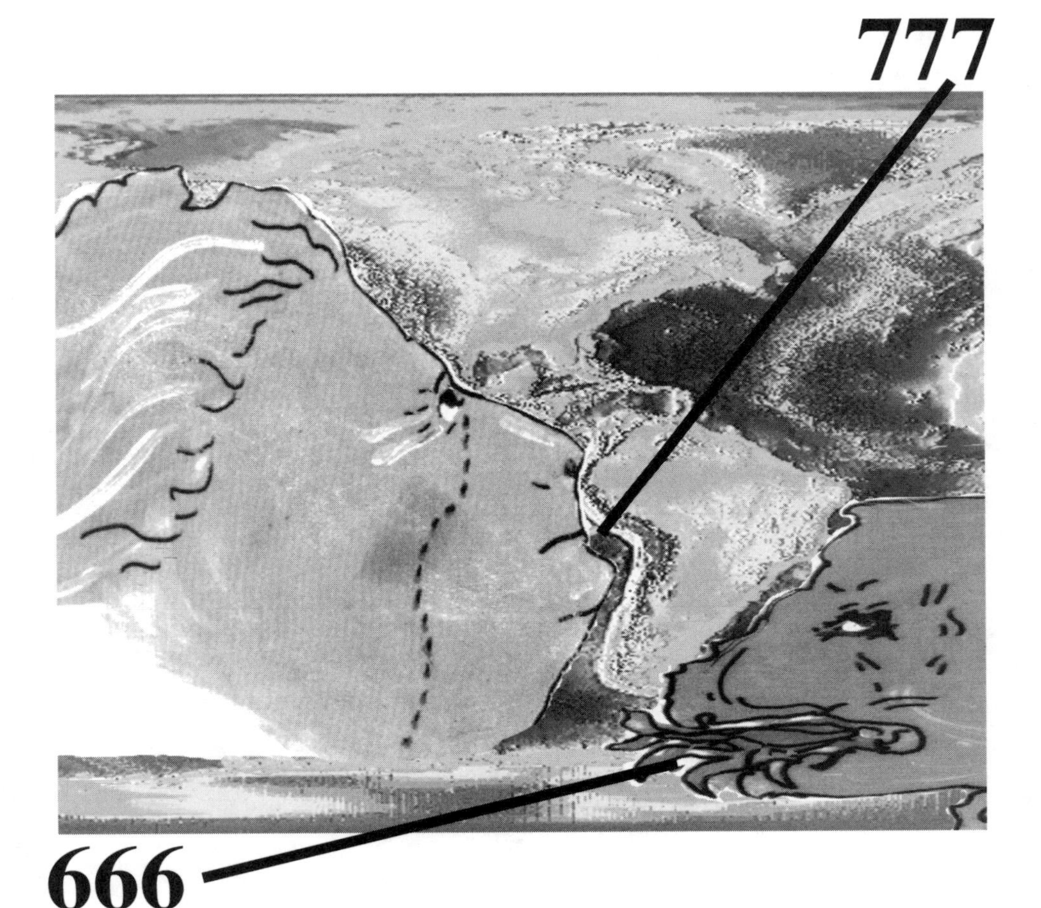

One of the problems with our world is that many people don't believe in the Devil. The Bible plainly supports the existence of this evil creature. It describes him as responsible for all the evil on this earth. He is known by Satan, the Adversary, Lucifer, Son of the Morning, Beelzebub (Matt. 12:24), Belial (2 Cor. 6:15), the Slanderer, the Devil and many other names.

THE NATURE OF SATAN

1. He is a creature. Ezekiel describes Satan as a cherub who was among the most powerful in all of Heaven. Later Satan descended to such a miserably low point that he is described as having then walked, "up and down in the midst of the stones of fire" (Ezek. 28:12-14).

2. <u>The Devil is a spirit being</u>. He spiritually hounds and works to sway all of humanity into continual discord. He is also known as the true spirit of confusion.

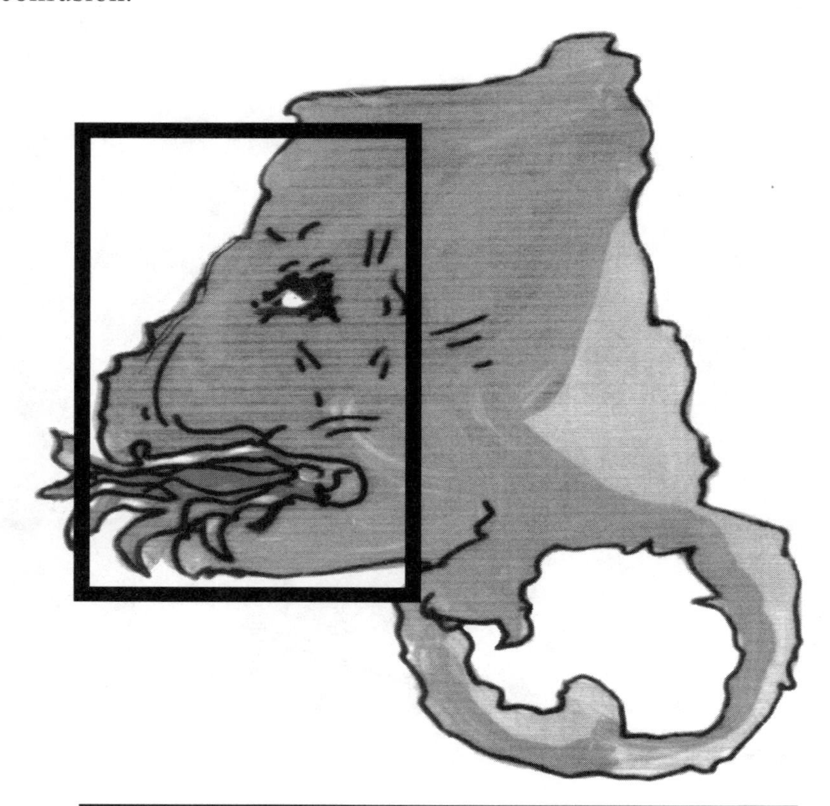

The area within the black square above represents the same part on this tracing that is emphasized on the map on the next page.

THE TITLES OF SATAN

1. The God of this Age (2 Cor. 4:4).
2. The Prince of this World (Jn. 12:31).
3. The Prince of the Power of the Air (Eph. 2:2).
4. The Accuser of the Brethren (Rev. 12:10).
5. The Tempter (1 Thess. 3:5).
6. The Evil One (1 Jn. 5:19, ASV).

This page highlights the “face” region of the Devil on the Rand McNally map of the South Atlantic Ocean Floor©

<u>**without an overlay**</u>.

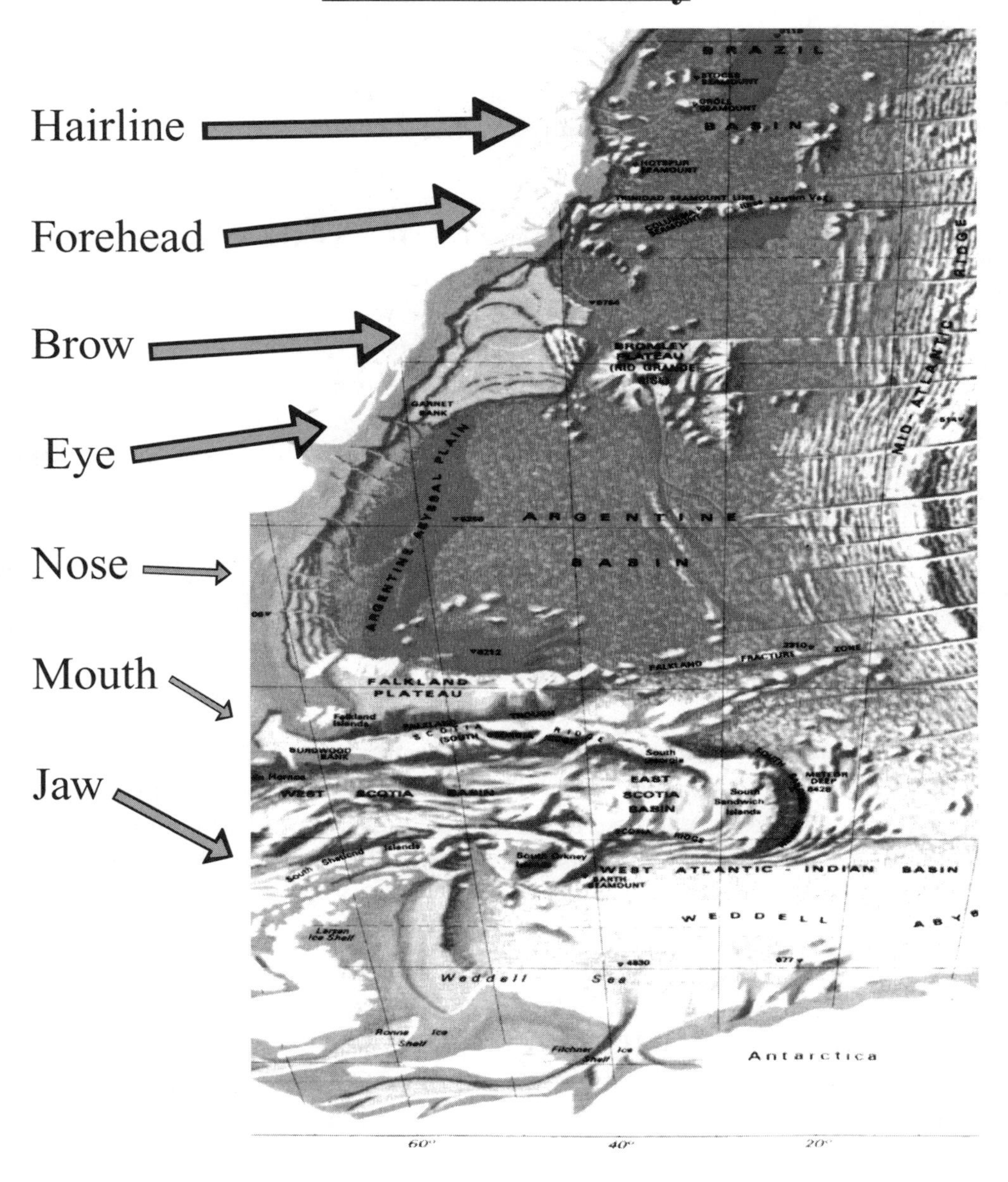

THE REPRESENTATION OF SATAN

The nature of Satan is not a mystery. His personality can be categorized distinctly. The Devil manifests himself throughout the Bible in three specific ways. In their order presented, he represents slippery cleverness, beastly ferocity and total fraud.

1. As a Serpent (Gen. 3:1, Rev. 12:9).
2. As a Dragon (Rev. 12.3).
3. As the Angel of light (2 Cor. 11:14).

THE PERSONALITY OF SATAN

1. He has the intellect to beguile and corrupt (2 Cor. 11:3).
2. He has angry emotions that desire war (Rev. 12:17).
3. He has an avarice will (2 Tim. 2:26).
4. He is morally responsible for his actions (Matt. 25:41).
5. He is in every way a person (Job 1).

THE TRAITS OF SATAN

1. He is a murderer (Jn. 8:44).
2. He is a liar (Jn. 8:44).
3. He has been a sinner from the beginning (1 Jn. 3:8).
4. He is an accuser (Rev. 12:10).
5. He is an adversary (1 Pet. 5:8).

Satan has limitations. He is neither omniscient nor infinite. He can be resisted by those who believe in Christ (Jam. 4:7). The Lord God doesn't give the Devil total free reign. He sometimes puts specific limitations on his evil powers (Job 1:12). With all of his power, Satan has no power without God's permission.

WHAT DID SATAN DO WRONG?

Before the Earth was ever created, Satan was one of the angels called Cherubim (Ezek. 28:14). He was very privileged and considered "full of wisdom, and perfect in beauty" (Ezek. 28:11). He had what anyone could ever want, but he wanted more. Then he did something very bad. Even though

God gave him power over all the other creatures in Heaven, he still wasn't satisfied. He wanted to dethrone God and take over. And to make matters even worse, Satan proclaimed his intentions to all of Heaven. Isaiah describes Satan's foolishness and punishment.

Isaiah 14:12-15
"How art thou fallen from heaven, O Lucifer, son of the morning! how art thou cut down to the ground, which didst weaken the nations! (v13) For thou hast said in thine heart, I will ascend into heaven, I will exalt my throne above the stars of God: I will sit also upon the mount of the congregation, in the sides of the north: (v14) I will ascend above the heights of the clouds; I will be like the most High. (v15) Yet thou shalt be brought down to hell, to the sides of the pit."

This middle-ages illustration called *The Mouth of Hell* presents the unbiblical belief that the Devil devours souls in Hell. From the *Kalendrier des Bergiers*.[2]

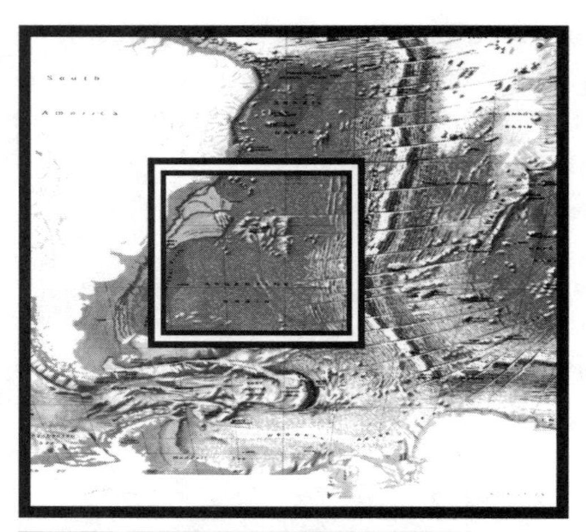

The Bromley Plateau and Rio Grande Rise

The Eye Of the Beast

This is an Enlargement of the Eye area of the Devil on the South Atlantic Ocean Floor.

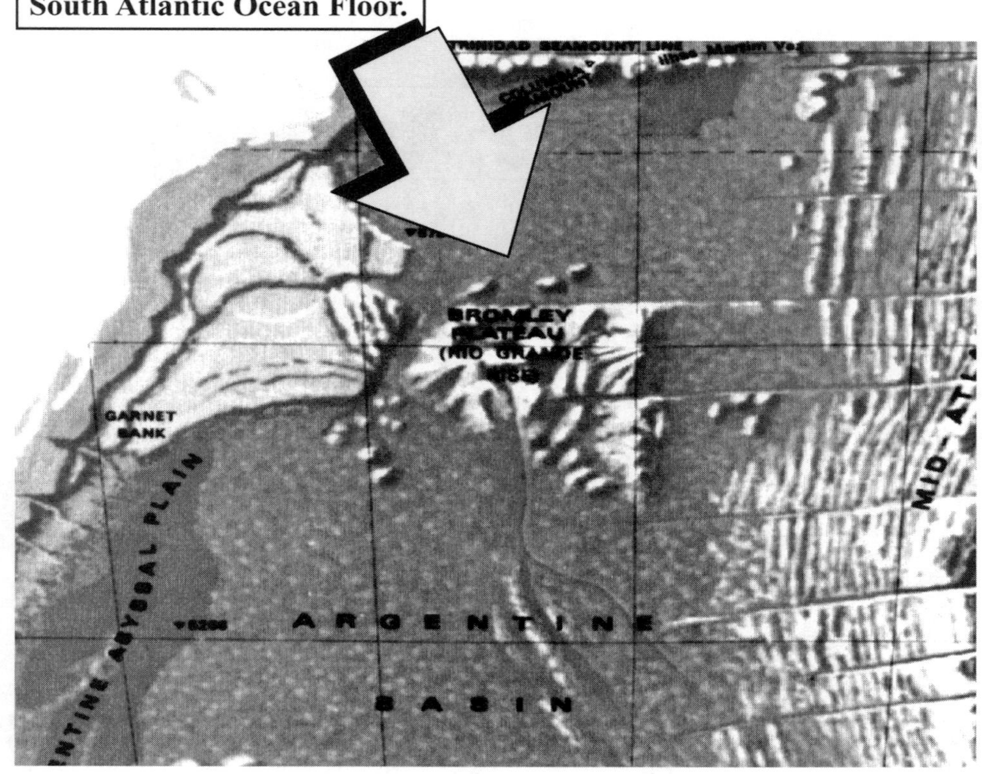

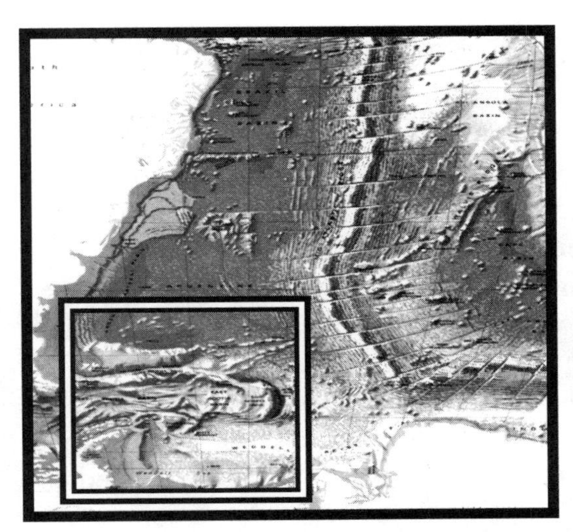

West Scotia Basin
and
East Scotia Basin

The Mouth Of the Beast

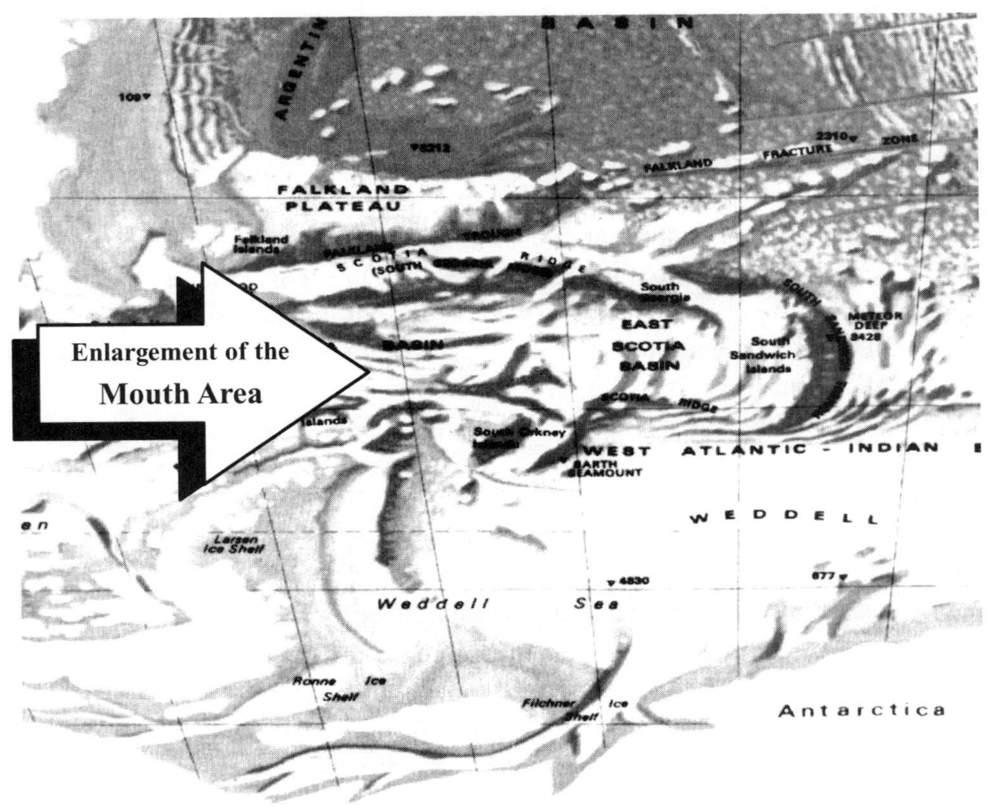

THE BEAST WILL BE CAST INTO THE LAKE OF FIRE

The main sin of the Devil is pride. He is characterized as attempting to be a counterfeit God, "like the Most High" (1 Tim. 3:6). God was so angered by Satan's actions that he cast him out of Heaven (Ezek. 28:16). Then he pronounced judgment on him in the Garden of Eden (Gen. 3:14-15) and continued that judgment at the cross (Jn. 12:31).

But the judgment and punishment of the Devil is again specified during the Tribulation Period (Rev. 12:13). At the beginning of the Millennium when Christ rules on Earth as King, Satan is confined in the Abyss (Rev. 20:2). At the end of one thousand years, Satan is briefly released but is still unrepentant. Then finally the Devil along with the beast and false prophet (Rev. 20:10) are cast into the lake of fire where they will be "tormented day and night forever and ever."

William Congreve wrote that "the Devil watches all opportunities." Shakespeare has Macbeth proclaim, "What, can the Devil speak true?" Isaac Watts warned that it is Satan who finds, "mischief for idle hands to do." Satan works continually to thwart the efforts of God and man. He is very clever but really not all that bright. Sometimes he is even stupid, like when he tried to tempt Christ (Matt. 4:1-11).

As a matter of fact, Satan used various people in an attempt to hurt the work of Jesus Christ (Matt. 2:16; Jn. 8:44; Matt. 16:23). The Devil even went so far as to possess the body of Judas so that this disciple could betray Jesus (Jn. 13:27). We know that Satan is the deceiver of all the nations (Rev. 20:3). As a matter of fact, he will gather them together for the battle of Armageddon (Rev. 16:13-14) which earmarks the end of the world.

Lucifer is always up to his old tricks and his main weapon is the simple lie. He uses lies to bind the minds of Christians (2 Cor. 4:4) and snatches the word of God from their hearts (Lk. 8:12). The Devil hates Christians and accuses them and slanders them before the very throne of God (Rev. 12:10). It is his pleasure to hinder the work of Godly people (1 Thess. 2:18) and he tempts them to lie (Acts 5:3).

The Devil loves to sabotage the work of God's people (1 Thess. 2:18). He uses demons to help him in his filthy deeds (Eph. 6:11-12). This persecutor (Rev. 2:10) sows tares among believers and tempts them to immorality (1 Cor. 7:5). Satan's army are demons which have various abilities. They can inflict diseases (Matt. 9:33; Lk. 13:11,16). They can possess humans (Matt. 4:24) and animals (Mk. 5:13) and even disseminate false doctrine (1 Tim. 4:1). The Bible makes it clear, every human being alive is constantly in danger of demonic influence.

Is this the man mentioned in the scripture below?

Isaiah 14:16-17

"They that see thee shall narrowly look upon thee, and consider thee, saying, ***Is this the man*** *that made the earth to tremble, that did shake kingdoms; (v17) That made the world as a wilderness, and destroyed the cities thereof; that opened not the house of his prisoners?"*

HOW CAN I PROTECT MYSELF?

After all this talk about the Devil and demons, it is understandable why some people might panic. The evidence seen on the South Atlantic Ocean Floor seems to prove the possibility of there being a Devil. Since this is so, how can you and I protect ourselves against the Devil and his demons? The best thing that anyone can do is become a repentant and Godly person who prays often. I believe that if you do this in the name of Jesus you please God, and more importantly really make things difficult for the Devil. Get your doctrine and understanding through frequent Bible study and fellowship with a local church you can trust.

This lifestyle is where the believer can confidently take a stand against Satan (Jam. 4:7). It is now possible to know why we must be on our guard against the Devil (1 Pet. 5:8). A holy life is still no guarantee that the Devil will stay away. God sometimes allows the Devil to bring misery to people that are completely innocent. An example of this would be the story of Job in the Bible.

NASA Views the

South Atlantic Ocean Floor *From Space*

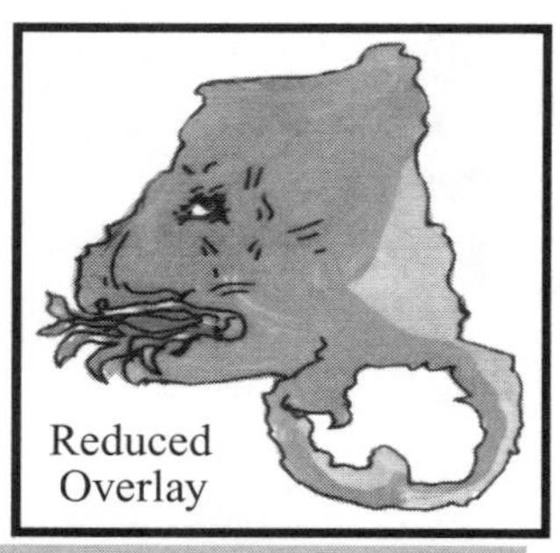

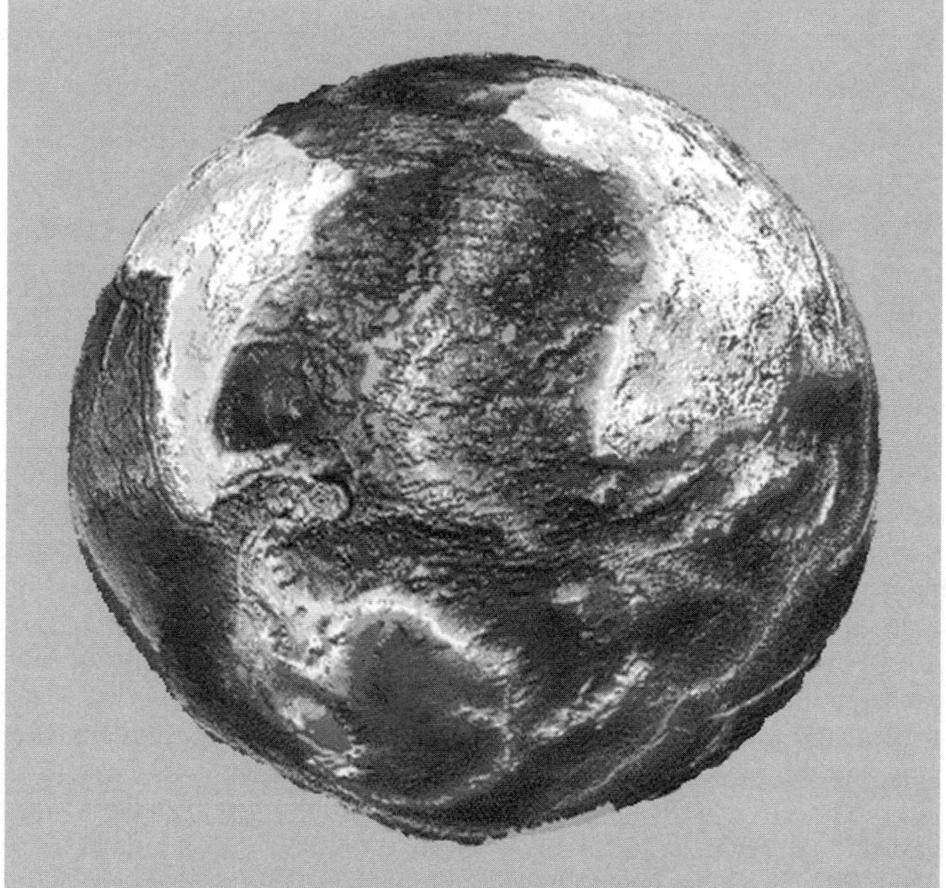

This global view, highlighting the South Atlantic Ocean Floor, is from the U.S. National Geophysical Data Center. The shading and contrast were chosen to give a natural look to the continents and oceans according to elevation. Major data sources include the U.S. Oceanographic Office, the U.S. Defense Mapping Agency and NASA.

From ancient times to the present, many people have believed that the location of Hell is at the center of the Earth. The Devil, having horns, a tail and a pitchfork is the eternal punisher. Human souls that are damned are to be made slaves in the Devil's fiery furnace. None of this mythology is found in the Bible and is considered pagan philosophy by most Christians, Jews and Muslims. Whatever Hell might look like is not as important as the uniformly agreed upon result of damnation: A permanent absence from the presence of God.

Could the Relationship Between These Two Persons Help Unlock the Mystery of the

Secret of Life?

These Two Figures Dominate the Topography of this Planet. The Evil Demon on the South Atlantic Ocean Floor is Full of Hate and Attacking the Weeping Man on the Pacific Ocean Floor.

Why were we born is life's ultimate question. Can the scenario of good versus evil shown on the topography of planet Earth be a window to truth? All living things are affected by good and evil. Are not the powers of good and evil the dictators of our every thought? Is it not easy to see that both separate influences are independent and can never meld? What else is this but an ultimate substantiation of moral law? This can only serve as a reminder. No man can serve two masters. Nobody can be loyal to both.

An Unmistakeable War Between *Good and Evil*

This is an unaltered section of a relief map of the surface of the Earth. It is from the Marine Geology and Geophysics Division of the NOAA National Geophysical Data Center. It was generated from digital data bases of land and sea-floor elevations on a 5-minute latitude/longitude grid. Assumed illumination is from the west; shading is computed as a function of the east-west slope of the surface with a nonlinear exaggeration favoring low-relief areas. According to NOAA, a Mercator projection was used for the world image, which spans 390 degrees of longitude from 270 degrees West around the world eastward to 120 degrees East; latitude coverage is (+) or (-) 80 degrees.

WEARING THE WHOLE ARMOR OF GOD

According to the Bible, sometimes God may even use Satan for beneficial purposes in the lives of his children. Even though we should rebuke the Devil at every turn, we should not speak of Satan with overt contempt (Jude 8-9). It is a waste of time and also because that is God's territory. Finally, there is probably no other section of the Bible that gives better information in regards to avoiding the Devil and his powers than this encouraging section from the book of Ephesians. The key here is to wear the whole armor of God.

Ephesians 6:11-14

"Put on the whole armor of God, that ye may be able to stand against the wiles of the Devil. (v12) For we wrestle not against flesh and blood, but against principalities, against powers, against the rulers of the darkness of this world, against spiritual wickedness in high places. (v13) Wherefore take unto you the whole armor of God, that ye may be able to withstand in the evil day, and having done all, to stand. (v14) Stand therefore, having your loins girt about with truth, and having on the breastplate of righteousness;

Ephesians 6:15-18

(v15) And your feet shod with the preparation of the gospel of peace; (v16) Above all, taking the shield of faith, wherewith ye shall be able to quench all the fiery darts of the wicked. (v17) And take the helmet of salvation, and the sword of the Spirit, which is the word of God: (v18) Praying always with all prayer and supplication in the Spirit, and watching there unto with all perseverance and supplication for all saints;"

Chapter 10

THE DRAGON

(On the North Atlantic and Arctic Ocean Floor)

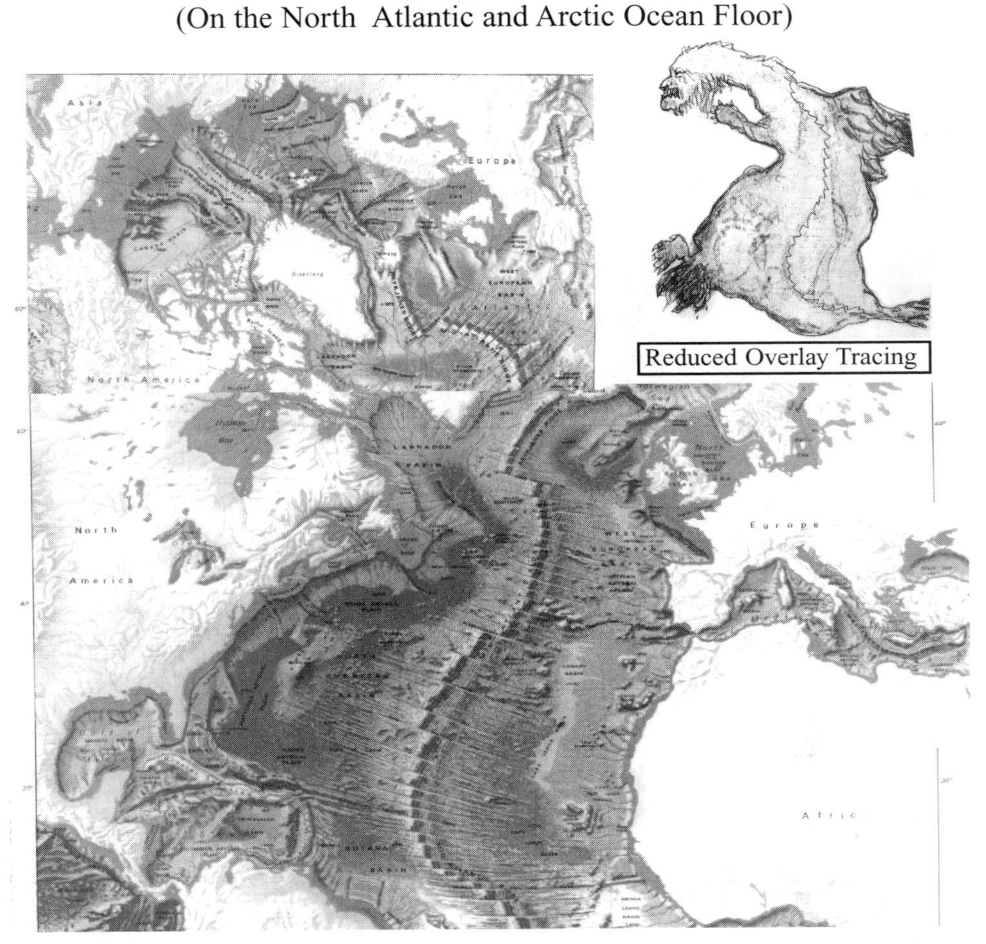

Reduced Overlay Tracing

Most people know that Godzilla is a myth and that dragons don't really exist today. Still the word "dragon" appears in the Bible no less than thirty-six times, both in the Old and New Testament.[1] The dragon, which is seen when aligning two Rand McNally© ocean floor maps, is in some ways the most fascinating image of them all. Here we connect topographic maps of the North Atlantic Ocean Floor and the Arctic Ocean Floor. The same criteria that we used with the Pacific and South Atlantic maps are also here.

This page shows a conceptualized border tracing of two topographic maps connected. We call this image:

The Dragon

This dragon, on the floor of the Arctic and North Atlantic *is blowing fire from his tail* into the face of the weeping man on the Pacific Ocean Floor.

These are the two ocean floor maps connected which make up the Dragon. The head is the Arctic and the body is the North Atlantic.

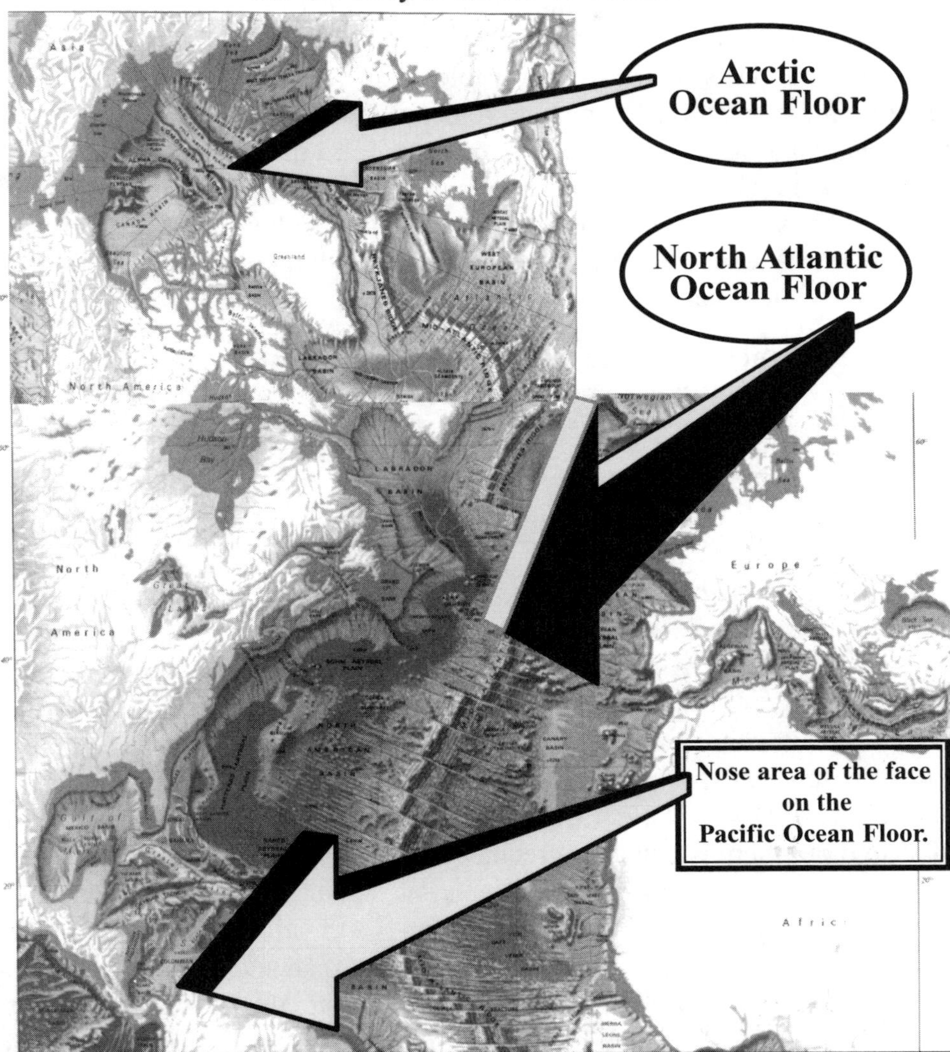

These are major sections of two topographic maps which are connected. They are the North Atlantic Ocean Floor and the Arctic Ocean Floor. Both maps are produced by the Rand McNally Company.©

TRACING WHERE THE WATER MEETS THE LAND

Of course, our border tracing is slightly embellished. But notice how it still fits our basic criteria: *Border tracing is only allowed where the water meets the land. Other than that, only the highest or deepest underwater mountain ranges are traced.* This is done to help identify the images traced. Still, the dragon is totally recognizable with no tracing at all. The actual topographic maps include much more detail than our tracing. It can be said that this image looks like a dragon or some kind of serpent. But, if it is a dragon, then it is important we know more about dragons.

The word "dragon" appears in the Bible in both the Old and New Testaments. A concise definition of the word "dragon" is offered by "An Expository Dictionary of New Testament Words." It reads:

> **DRAGON:** Denoted a mythical monster, a dragon; also a large serpent, so called because of its keen power of sight (from a root derk, signifying to see). Twelve times in the Apocalypse, it is used of the Devil.[2]

My 1982 letter from Rabbi Alan Lachtman (see page 63) had an interesting last paragraph. He wrote, *"I find equally amazing, the configuration of a dragon in the sonogram of the North Atlantic Ocean, which echoes the Mesopotamian tradition that the earth and its heaven were formed with the aid of the sea dragon, Tiamat."* [3]

Tiamat was the dragon of the "salt waters." The legend states that Tiamat was unconquered. She had a brood of monsters attacking the gods who were unable to prevail against her.[4] Some legends tell of Tiamat working on the side of God and others against God. Dragons are not always thought of as evil but there is general agreement about one thing, you don't want to be around one when he is mad.

The dragon seen in the tracing of the North Atlantic Ocean Floor and Arctic Ocean Floor is ferocious. He is mean and mad and in the process of a wild and angry fit. He is blowing fire out his tail into the tears of the Pacific. His teeth are prominent and he is growling. He is helping the Devil on the South Atlantic Ocean Floor attack the weeping man on the Pacific Ocean Floor. This is obviously not a good dragon but worthy of one name given to the Devil, "The Beast."

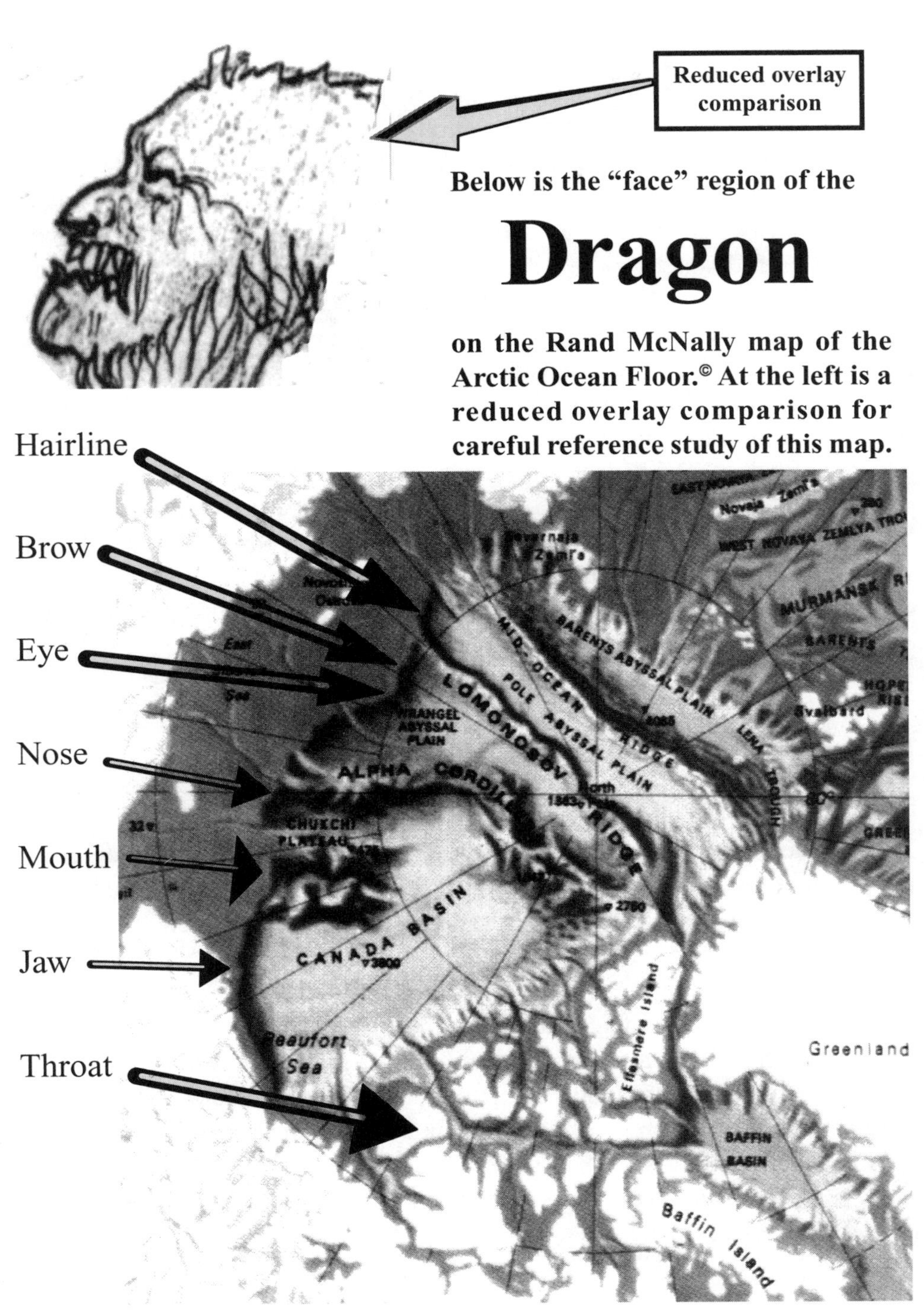
Reduced overlay comparison
Below is the "face" region of the
Dragon
on the Rand McNally map of the Arctic Ocean Floor.© At the left is a reduced overlay comparison for careful reference study of this map.
Hairline
Brow
Eye
Nose
Mouth
Jaw
Throat
WRANGEL ABYSSAL PLAIN
LOMONOSOV RIDGE
MID-OCEAN RIDGE
POLE ABYSSAL PLAIN
BARENTS ABYSSAL PLAIN
ALPHA CORDILLERA
CHUKCHI PLATEAU
CANADA BASIN
LENA TROUGH
MURMANSK
WEST NOVAYA ZEMLYA TROUGH
Svalbard
Beaufort Sea
Ellesmere Island
Greenland
BAFFIN BASIN
Baffin Island

THE DRAGON AS SYMBOLISM

I find it fascinating that the dragon traced with these two maps appears to seamlessly flow out of the head or mind of the Devil on the South Atlantic Ocean Floor. One example showing this relationship between the North Atlantic and South Atlantic can be seen on page 153 of this book. It is as if this ferocious creature is in actuality the measure of Satan's thoughts. For the most part, this is how the Bible refers to the dragon. The dragon is seen as an influence that does the Devil's bidding and even as a person that is simultaneously Satan himself. A study of the dragon throughout the world's mythology could be quite extensive. Still, it would not bring us much closer to understanding the symbolism represented by the dragon on the North Atlantic Ocean Floor and the Arctic Ocean Floor. To do this we will need to take a closer look at what the Bible says about dragons.

The first time we see the word "dragon" appear in the Bible is in the book of Deuteronomy. God wanted Israel to know that he is absolutely sovereign and that he will ultimately judge Israel's enemies and restore his chosen people. The warning to Israel is to be aware of the evils of their enemies. The King James Version (KJV) reads:

Deuteronomy 32:33
"Their wine is the poison of dragons,
and the cruel venom of asps."

The New American Standard (NASV) replaces the word "dragon" for "serpent," but the meaning of the warning remains the same. Later in the book of Nehemiah (2:13), we see the term "dragon-well" used in both the KJV and the NASV but we are not sure what this term means precisely. Possibly it means a snake-pit or garbage dump which was located near the southwest Valley Gate of Jerusalem.[5]

When Job (30:39) (KJV) is lamenting in agony, he refers to himself as the "brother to dragons, and a companion to owls." Here the word "dragons" is replaced by the word "jackals" in the NASV, which may be the more appropriate term..

The book of Psalm refers to dragons on four occasions.The first (44:19) is likely referring to a jackal but the second is certainly in reference to a sea monster of some kind. First, there is a reference to God "dividing the seas" by his strength, resulting in his breaking the heads of dragons. Notice that the writer, King David, begins by stating that salvation is being worked out by God from a specific location. The passage reads that salvation occurs "in the midst of the earth." Then he refers to "dividing the sea." Could this be the use of a Hebrew double meaning. Could this also be a possible reference to sea-floor spreading or Continental Drift? We can only speculate on this. The section reads:

Psalms 74:12
"For God my King of old, working salvation
in the midst of the earth. (v13) Thou
didst divide the sea by thy strength:
thou brakest the heads of the
dragons in the waters."

In Psalms 91:13 the term "dragon" is used by the KJV and "serpent" by the NASV. In Psalms 148:7, David in adoration and praise to God orders that they, "Praise the LORD from the earth, ye dragons, and all deeps...." The first reference in the book of Isaiah to dragons (13:22) is probably referring to jackals but with the second occurrence, Isaiah is definitely speaking of a dragon that is somewhere in the sea. References that are similar to this appear mostly in the Old Testement. Speaking of God's final judgment, Isaiah writes:

Isaiah 27:1
"In that day the LORD with his sore and great and strong
sword shall punish leviathan the piercing serpent,
even leviathan that crooked serpent; and he
shall slay the dragon that is in the sea."

Twice later (35:7 and 43:20) the term is again used to describe a jackal. But in the 51st chapter, we see the word "dragon" used as a great sea monster and also as a metaphor referring to the oppression of Egypt on the Jews. Here the word "Rahab" is the mythological chaos monster equated with Egypt.[6]

Isaiah 51:9
"Awake, awake, put on strength, O arm of the LORD; awake, as in the ancient days, in the generations of old. Art thou not it that hath cut Rahab and wounded the dragon? (v10) Art thou not it which hath dried the sea, the waters of the great deep; that hath made the depths of the sea a way for the ransomed to pass over?"

The weeping prophet Jeremiah uses the term "dragon" on six occasions (9:11, 10:22, 14:6, 49:33, 51:34, 51:37). Five of them really mean jackal, but one is certainly speaking of some kind of monster. The NASV does translate it as "a monster." Here the king of Babylon is symbolically compared to in the KJV as a dragon.

Jeremiah 51:34
"Nebuchadnezzar the king of Babylon hath devoured me, he hath crushed me, he hath made me an empty vessel, he hath swallowed me up like a dragon, he hath filled his belly with my dedicates,he hath cast me out."

Ezekiel uses the word "dragon" only once, similarly to that seen in Isaiah (27:1) as a metaphor for Egypt and its pharaoh. Micah (1:8) and Malachi (1:3) are shown as "dragons" in the KJV but are more appropriately translated as jackals in the NASV.

LEVIATHAN THE DRAGON

There is another Old Testament term that can also mean a dragon. It is the word "Leviathan." It can mean a crocodile or other large sea-monster but also as a parallel to the constellation of the Dragon and even as a symbol for Babylon.[7] Job (41:1) probably means crocodile when he speaks of Egypt as Leviathan as does David in Psalms 74:14, although David's reference in Psalms 104:26, may be of a larger sea-monster. Sometimes it describes nations. When Isaiah uses the term "Leviathan" (27:1) it symbolizes the enemies of God.[8]

Although the Bible says nothing about dragons or serpents helping the Devil torment humans that are damned, such beliefs are part of Christian tradition. The above 1692 drawing depicts the torment of a sinner in Hell. It is from Father G.B. Manni's, *The Eternal Prison of Hell for the Hard-Hearted Sinner.*[9]

These references to "Leviathan" or a "dragon" are symbolic of the evil found in man turning away from God through his more carnal nature. Author Manly Hall elaborates, "According to many scattered fragments extant, man's lower nature was symbolized by a tremendous, awkward creature resembling a great sea-serpent, or dragon called Leviathan."[10]

NEW TESTAMENT REFERENCES

All of the references to a dragon in the New Testament occur in the book of Revelation. The first eight (12:3, 4, 7, 8, 9, 13, 16 and 17) each refer to the same dragon. It begins with Saint John attempting to describe a "great red dragon" (v3) seen in his vision. It is similar to that creature described by Daniel (7:7, 20:23-25) but called a "beast." A Bible commentary edited by Walter A. Elwell adds, "As in Daniel, the dragon represents evil political powers on earth that persecute and attempt to torment the church, but also their cosmic evil leader, Satan. He is described as red perhaps to symbolize his murderous character."[11]

INTO THE EARTH?

Christian tradition has St. George, St. Margaret and St. Michael slaying dragons. The battle of the Archangel Michael fighting a dragon is also told about in the Bible. It occurs when God has the Devil (the accuser) and the dragon (who does his bidding) cast out and thrown down to Earth. We see this in:

Revelation 12:7, 8 & 9
"And there was war in heaven: Michael and his angels fought against the dragon; and the dragon fought and his angels, (v8) And prevailed not; neither was their place found anymore in heaven. (v9) And the great dragon was cast out, that old serpent, called the Devil, and Satan, which deceiveth the whole world: he was cast out into the earth, and his angels were cast out with him."

It is interesting that the above passage states that the dragon was cast out "into the earth." This is especially fascinating when we consider the fact that there is a giant (imprint) of a dragon on our oceans floor. It is almost as if John saw this imprint himself. The other passages in Revelation (13:2, 4, 11, 16:13, and 20:2) using the term "dragon" even further clarifies the true identity of the dragon. He is (20:2) none other than the one and only, you know who, the Devil himself. But even the Devil needs a frontman, so here in disguise, he is the dragon.

Chapter 11

THE TWO ANGEL TYPE CREATURES

Are There Angels?

THE INDIAN OCEAN FLOOR

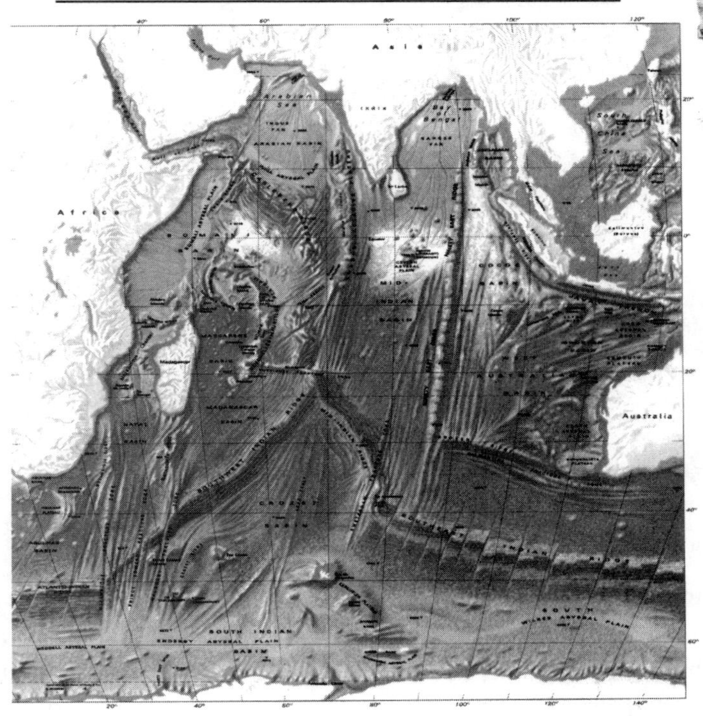

The topographic map of the Indian Ocean Floor shows what could be identified as two cherubs or angel type creatures. Although the images may not seem as clear to the first time observer as the others we've reviewed, our guess is based on the substantial evidence we've gathered so far. The figure that looks like he is holding a stick or sword in his right hand, has far more detail than the second figure which is not much more than a shadow.

It is not difficult to see the large-bellied, thick-thighed creature with his right arm extended and holding some type of long object. Even his right eye has tracing definition in the areas of placement and emotion. As a matter of fact, only his *right eye* is shown.

The second figure is the only one of all the ocean floor images (based on our border tracing criteria) that is not easily decipherable. Whatever this figure is, it's not facing us, and we see no eye. My guess is that since we see no eye anywhere on this particular image, then it must have its back to us. If anyone has any better suggestions, I'd like to know.

The interaction shows that the creature with the sword is pulling the Devil away from the face on the Pacific Ocean Floor. This is more apparent when viewing all the ocean floor maps (see page 153) and seeing them connected on one page. The eye of the creature with the sword is at Indus Canyon, his right arm is at the Gulf of Aden and his sword is the Red Sea Rift. One reason we refer to this creature as a possible cherub is due to a description we see in scripture. It happens when Adam and Eve are exiting the Garden of Eden in shame.

Since only a shadow of the second creature is seen, we call this figure an angel for convenience sake and because it fits the overall story line presented. But the first image is a key figure in this drama. He looks not unlike a character out of a novel by J.R.R.Tolkien or even like those "creature" type cherubs (with a flaming sword) described as guardians of the "Tree of Life" in the book of Genesis (3:24).

The existence of angels is a major theme in the Bible. Angels are instrumental figures at the beginning and the end of the world. They are referred to in at least thirty-four books of the Bible. The word "angel" occurs in scripture about two-hundred and seventy-five times. Christ knew of and taught about the existence of angels (Matt. 18:10, 26:53).

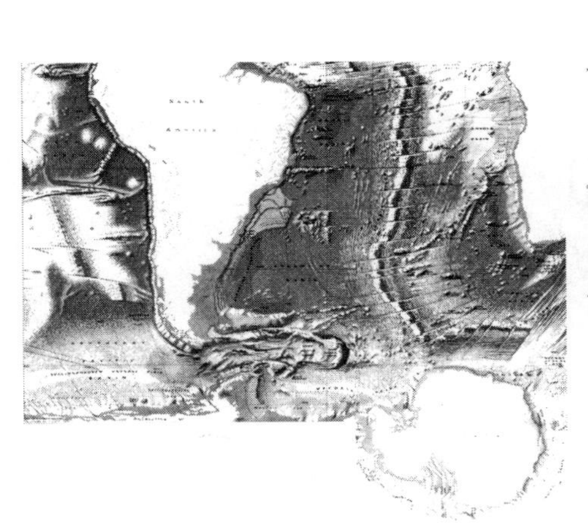

Genesis 3:24
"So he drove out the man; and he placed at the east of the garden of Eden Cherubims, and a flaming sword which turned every way, to keep the way of the tree of life."

We see the creation of angels in the book of Colossians (1:16). They were created before the beginning of the world (Job 38:6-7). Angels were created in a state of holiness (Jude 6). They are intelligent (1 Pet. 1:12), emotional (Lk. 2:13), and have individual will (Jude 6). Angels are spirit beings (Heb. 1:14) and do not have the ability to reproduce after their own kind (Mk. 12:25). Angels have great power (2 Pet. 2:11). They are distinct from human beings (Psa. 8:4-5) and they do not die (Lk. 20:36).

This page shows a conceptualized border tracing of the topographic map of the Indian Ocean. Could they be

Angels?

Compare this tracing with the next page. The angel with his sword seems to be pulling the Devil away from the throat of the weeping man on the Pacific Ocean Floor. An example of this relationship is on page 153 of this book.

His sword is the Red Sea Rift, his right arm is the Gulf of Aden and his right eye is the Indus Canyon. Can you guess the identity of the figure next to him? We are not sure.

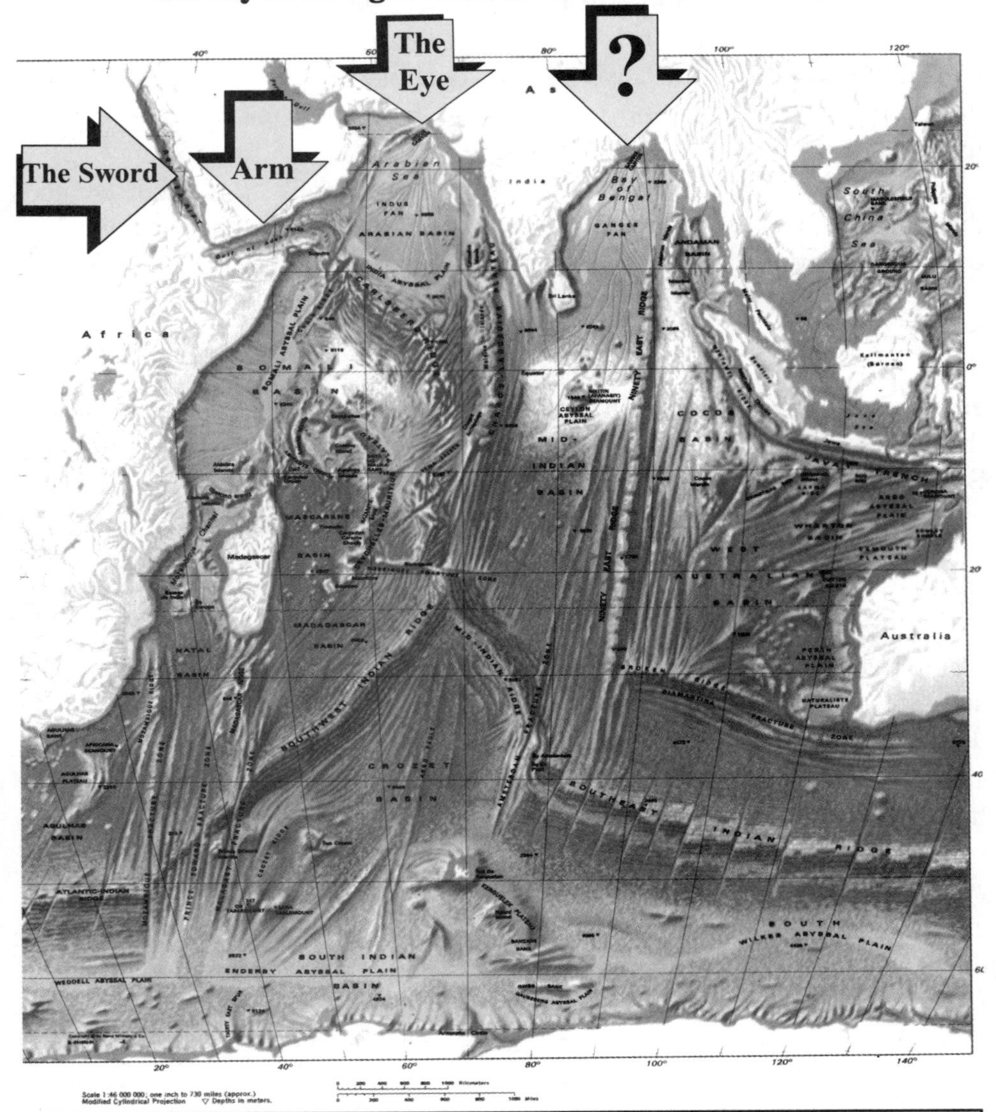

This is a topographic map of the Indian Ocean Floor. It is produced by the Rand McNally Company.© When studied carefully, it can be most fascinating.

ARE THERE ANGELS?

The idea that the large figure traced from the Indian Ocean Floor may represent an angel with a flaming sword is not without some merit. The first angels mentioned in the Bible are described in just such a way. They were called "Cherubim." It was their job to guard the Garden of Eden which housed the "Tree of Life." We see this documented in: (Gen. 3:24) *"...he placed at the east of the Garden of Eden, Cherubims, and a flaming sword which turned every way....*"

But exactly what are angels? Where did they come from? Who created them and what functions do they serve, if any? Although space prohibits an entire thesis on the subject here, some brief explanations are in order.

According to the Bible, angels were created by God (Col. 1:16). They were created before the beginning of the world (Job 38:6-7). The Bible states that angels were made in a state of holiness (Jude 6). In many ways angels are similar to humans, so much so that sometimes they are discribed as exactly that. Most of the time angels in the Bible are described as looking like men. Sometimes, they are described as animal or bird-like with multiple eyes and even wings. Every angel has their own identity and personality. They possess intelligence (1 Pet. 1:12), express emotions (Lk. 2:13) and they each have their own will (Jude 6).

It might seem confusing to some people but angels appear throughout scripture in both human and animal form. Angels seem to come in all shapes and sizes. There are good angels and bad angels. Also there has been much debate throughout history concerning the power of angels and even their size. Most of us have heard the expression, "How many angels can fit on the head of a pin?" At first glance this seems like a silly question. But such speculation was an important part of priestly studies. If an angel can be as big as trees and also the size of men, how small might they be? Could they even be molecular or atomic in size? If so, then millions of angels can fit on the head of a pin with no problem.

Our speculation is that the topographic map of the Indian Ocean Floor represents the image of at least one and possibly two angels. One reason for this is because the figure with the stick or sword in his hand seems to be pulling the Devil off the throat, of the face, on the Pacific Ocean Floor. Also, like the face on the Pacific Ocean Floor, all we see is his *right* eye.

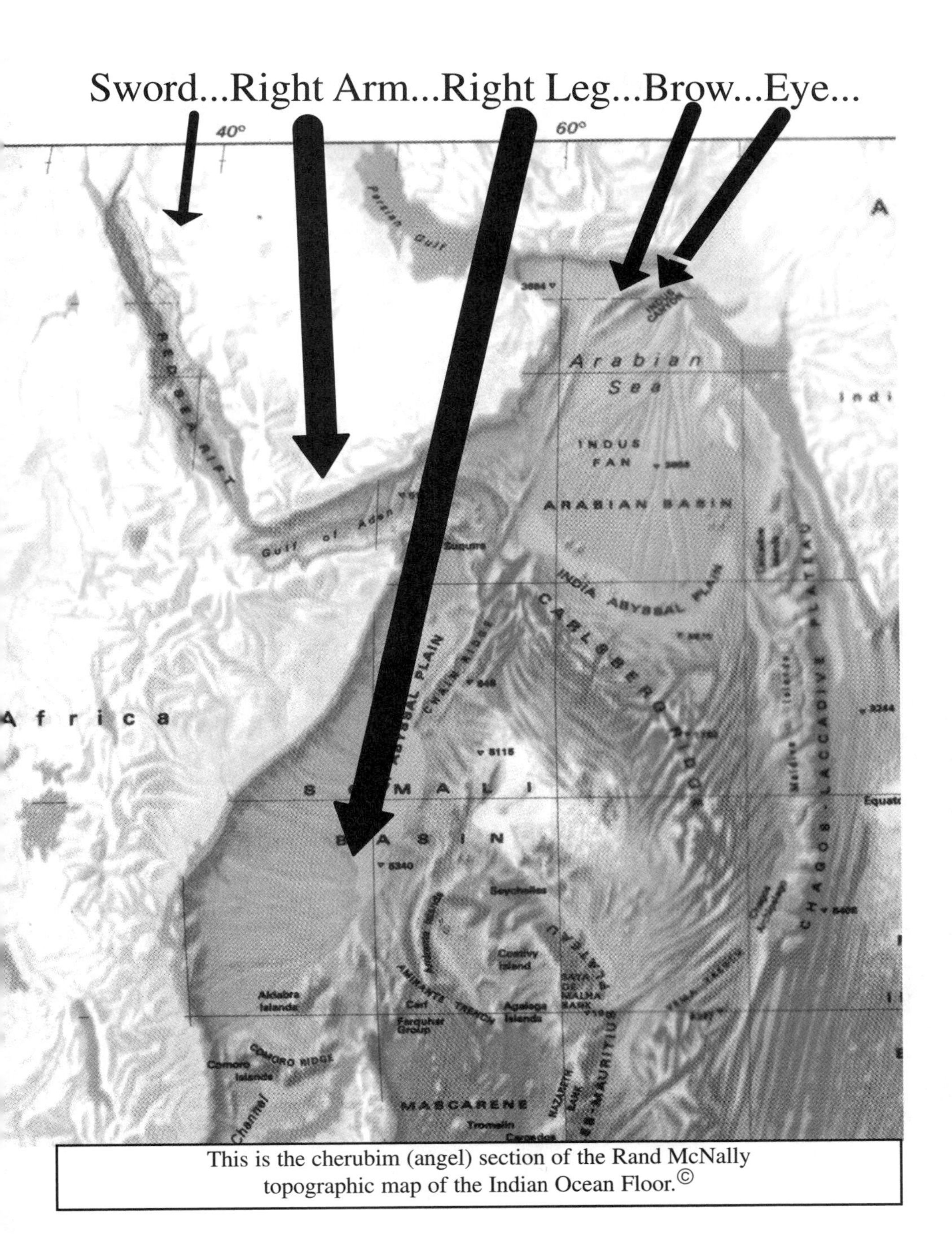

This is the cherubim (angel) section of the Rand McNally topographic map of the Indian Ocean Floor.©

The Flaming Sword Area Near the Indian Ocean Floor is Called the RED SEA RIFT and His Arm is The GULF OF ADEN.

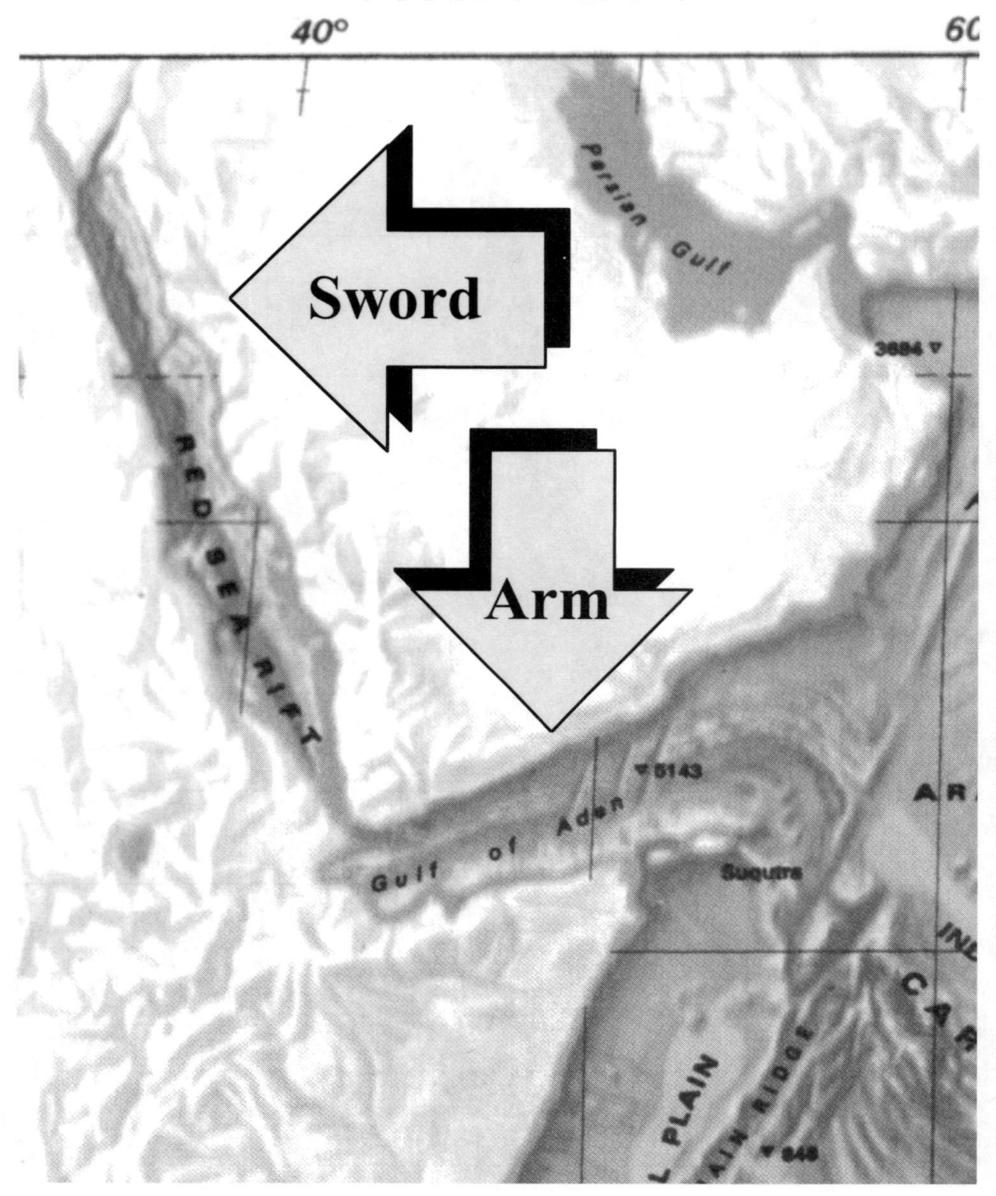

Here We See that His Eye is Near the INDUS CANYON. His Bearded Face is the INDUS FAN and ARABIAN BASIN Sections of the ARABIAN SEA.

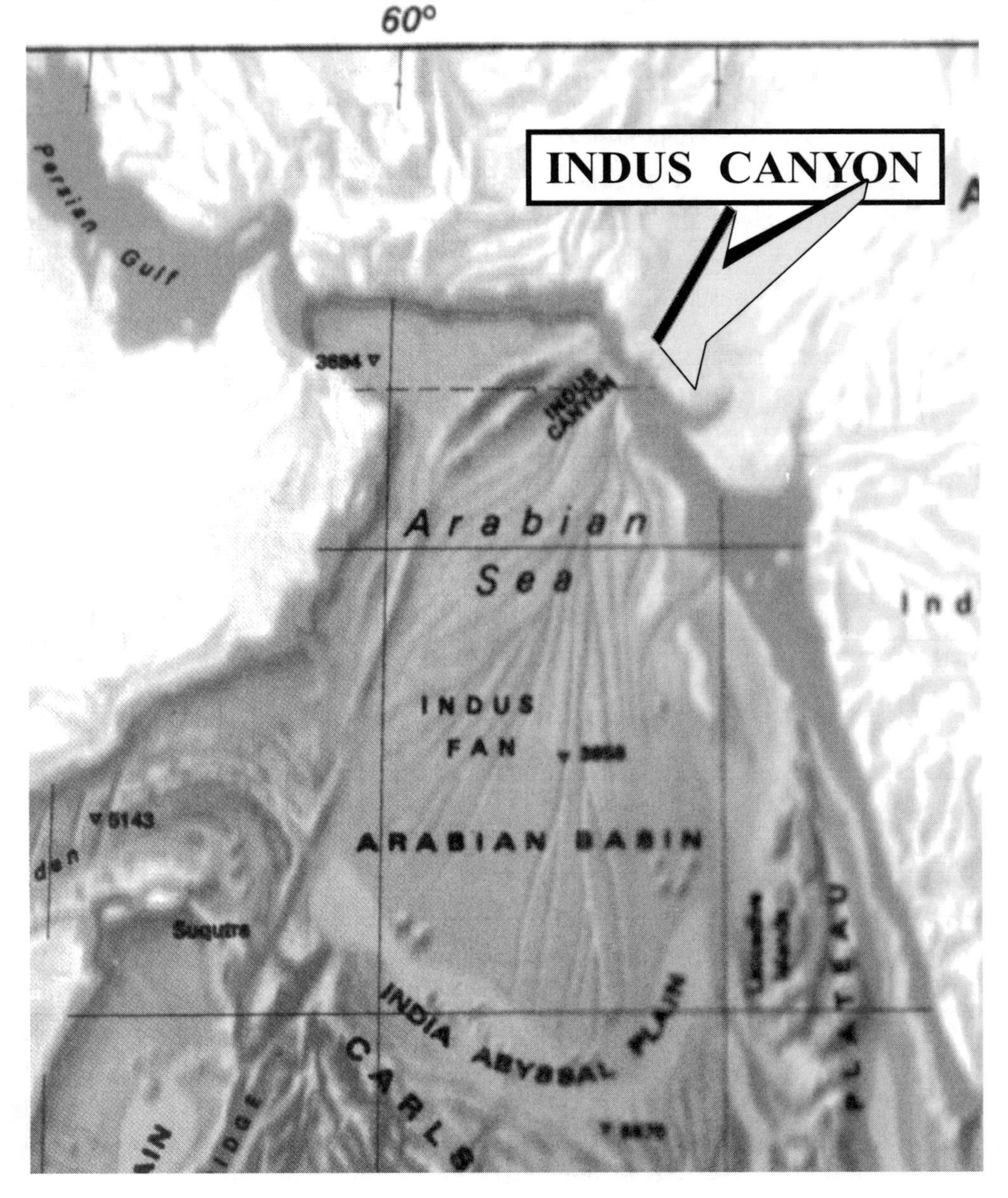

Angels come in many shapes and sizes but all have some similarities. Even though angels are not able to reproduce offspring of their own kind (Mk. 12:25), there is some evidence that they can reproduce unique offspring when mating with humans. Apparently, intermarrying between angels and "the daughters of men" resulted (Gen. 6:1-4) in a race of giants called Nephilim. Another interpretation of this passage is that it refers to intermarrying between the lineage of Sethites and the Cainites. The problem with this interpretation is that all other references to "sons of God" in the Old Testament describe angels.[1] In any case, God was so displeased with this sin, that he shortened the life-span of human beings to one hundred and twenty years.

Genesis 6:1-4
"And it came to pass, when men began to multiply on the face of the earth, and daughters were born unto them, (v2) That the sons of God saw the daughters of men that they were fair; and they took them wives of all which they chose. (v3) And the LORD said, My spirit shall not always strive with man, for that he also is flesh: yet his days shall be an hundred and twenty years. (v4) There were giants in the earth in those days; and also after that, when the sons of God came in unto the daughters of men, and they bare children to them, the same became mighty men which were of old, men of renown."

Angels in scripture are shown in human, animal and birdlike form, or a mixture of all three. Abraham saw angels as men (Gen. 18:1-2) and Zechariah saw them as women with wings like a stork. There is some argument whether these passages in Zechariah actually describe good angels of God from Heaven or those wicked ones banished to Shinar (Gen. 11:2 and Rev. 17:3-5). Most Bible stories about angels describe them as male. It is only in Zechariah that we can see evidence that some angels are also female.

This is among the most recent computer generated images based on scientific data and focusing on the

Indian Ocean Floor

Overlay Comparison

This global view highlighting the angels on the Indian Ocean Floor is from the U.S. National Geophysical Data Center. The shading and contrast were chosen to give a natural look to the continents and oceans according to elevation. Major sources include the U.S. Oceanographic Office, the U.S. Defense Mapping Agency and NASA.

Zechariah 5:9-11
"Then lifted I up mine eyes, and looked, and, behold, there came out two women, and the wind was in their wings; for they had wings like the wings of a stork: and they lifted up the ephah between the earth and the heaven. (v10) Then said I to the angel that talked with me, Whither do these bear the ephah? (v11) And he said unto me, To build it an house in the land of Shinar: and it shall be established, and set there upon her own base."

It is sometimes surprising how much we can learn about angels by studying the Bible. For one thing angels do not die (Lk. 20:36). Though they can resemble human form, they are intrinsically different and distinct from human beings. Some angels are servants of God and carry with them great power (2 Pet. 2:11).

According to the Bible, angelic organization has definite structure. The highest rank of angel is the Archangel, such as Michael (Jude 9). Some angels hold the rank of "Chief Princes" (Dan.10:13). Others are "Ruling Angels" (Eph. 3:10). Also there are of course, Guardian Angels (Heb. 1:14; Matt. 18:10), Seraphim (Isa. 6:1-3), Cherubim (Gen. 3:22-24), and "Angels of the Elect" (1 Tim. 5:21).

GOD'S SECRET ANGELS

The structure and identification of angels is so respected that its study is a part of most seminary curriculum. These helpers of God are an integral part of his master plan. Still we must be very careful to limit our descriptions of how angels might or should appear. One reference of angels in the bible simply calls them "strangers." We are told to be good to strangers as a practice because they might be "angels unawares" or angels without us (or even them) knowing it.

Hebrews 13:2
"Be not forgetful to entertain strangers: for thereby some have entertained angels unawares."

Angels are messengers and workers for God. They predicted the birth of Jesus Christ (Lk. 1:26) and announced his birth (Lk. 2:13). They not only predicted his birth but even protected the baby Jesus (Matt. 2:13). When Christ needed strength after being tempted by the Devil, it was the angels that ministered to him. Later when Christ needed help, they were ready to defend him (Matt. 26:53). In the garden located at Gethsemane, Jesus felt very weak. He only wanted to do the will of God. He was sweating blood while praying and God sent him angels (Lk. 22:43).

Angels do not pretend (with the exception of Satan) to have the power of God or even desire it. But Satan did take one third of the angels with him when he was cast out of Heaven. Even so, God sends his heavenly angels as ministering spirits.

Hebrews 1:13-14
"But to which of the angels said he at any time, Sit on my right hand, until I make thine enemies thy footstool? (v14) Are they not all ministering spirits, sent forth to minister for them who shall be heirs of salvation?"

Muslims believe that angels accompanied Muhammad on his night journey to heaven and also that the angel Gabriel appeared to the prophet in the cave on Mt. Hira to announce his ministry.[2] Christians and Jews naturally delineate between their views of angels and those of other cultures. As a Christian, I also am most comfortable looking to the Holy Bible for information about angels. God has many special angels that watch over Christians in their daily walk through life.

There seems to be no limit to the good that God's angels can do. For believers, they are involved in answering prayer (Acts 12:7). They also give encouragement in time of danger (Acts 27:23-24). They do everything they can to help God's workers evangelize (Lk. 15:10; Acts 8:26). Angels are not only observers of human experience (Tim. 5:21; I Cor. 4:9), but they also care for the righteous at death (Lk. 16:22; Jude 9).

My goal here is not to offer a complete teaching on angels in particular. All that is necessary is the cursory review of a Bible concordance and anyone can dive into a more thorough study. Also many excellent books on the subject of angels are available at most book stores and public libraries. But when considering the identification of the creature seen on the Indian Ocean Floor, the idea of it being an angel makes sense. He guards the welfare of God, just as angels work on behalf of humans.

Psalm 91:11-12
"For he shall give his angels charge over thee, to keep thee in all thy ways. They shall bear thee up in their hands, lest thou dash thy foot against a stone."

When we connect the activities of all of the four major ocean floor images, we see that they link. There is an interaction occurring that is recognizable. We can see that the weeping man on the Pacific Ocean Floor is being attacked by the Devil and a dragon. Attempting to pull the Devil off of the throat of the weeping man on the Pacific Ocean Floor is a large creature holding a big sword in his raised right hand. No one can say for certain that this creature depicted, represents an angel, but in time, when all is revealed, then we will know for sure.

I Corinthians 13:12
"For now we see through a glass, darkly; but then face to face: now I know in part; but then shall I know even as also I am known."

Chapter 12

ALL THE OCEANS IN THE WORLD

How Do They Relate To Each Other?

The answer to the question presented in this chapter has everything to do with substantiating the validity of the face on the Pacific Ocean Floor. Concerning this topic, reality has to be more than subjective. One person can see a face in the clouds where another person may not. Very few people refuse to acknowledge the face on the Pacific Ocean Floor as real. But people who see the face on the Pacific Ocean Floor with unbelief aren't always closed minded. They simply want more evidence. That is the main reason the question in the title of this chapter is so important.

PROVE IT

It is inevitable that many first-time observers approach my discovery with hesitance. I am continually asked to prove my theory with more and more additional evidence. I have learned to respond with patience. Some first-time observers can actually be quite vocal, even to the point of later embarrassment to themselves. This is because, at first, this whole concept is so unbelievable that it escapes our common reality. Every new discovery comes with some fear of the unknown. If God, whose dimension is outside our understanding is directly communicating with all of humanity in a three-dimensional way, then what is next? The answer is too complex for me, but the question occurs to everyone. The possibilities are endless.

FINDING A WRITTEN LETTER FROM GOD

What if scholars stumbled upon a seven-page letter, supposedly authored by God? Undoubtedly, they would do everything possible to test its authenticity. They would look at the style of writing, the language, the grammar, the paper and the age of the document. But God did not choose to use paper or writing or ink or anything else to scribe his message to humanity. He did it in a manner that only God could. He etched his written letter on mountains of stone, and he made his message non-erasable by having it cover the entire earth.

Creating a message *with pictures* that validate the Bible was an awesome choice for God to make. By writing that message on the entire planet, God made it possible for research to come from several unrelated fields at once. Especially in our day, the story told through the images on our oceans floor, encourages serious Bible study in a powerful way. Topographic map-makers will have a new zeal with their every measurement. Data banks will make possible an exchange and comparison of the latest information.

Every new piece of evidence, every new link, strengthens this most wonderful and exciting discovery. A few scientists and others may try to debunk this phenomenon but this is unimportant if the common layperson can benefit. That is the true advantage of these images. They cannot be erased by anyone but God.

This is not the kind of discovery that is easily hidden. Arguments and debates may go in many different directions but words won't change the evidence. Seventy percent of our planet comprises four basic sets of images that link logically. In a way, by looking at it we can see the genius of God's power. At first, like "Pandora's Box," it seems so innocent that it simply invites pause and contemplation. Then something happens that we do not expect. These images begin to make sense.

HOW DO THEY RELATE?

To some people, the massive weeping face on the Pacific Ocean Floor, covering almost one-half of our planet, might seem like a fluke. If this was the only ocean floor image, such doubt might be understandable. But if more images can be found, and they have, then it only adds to the beauty of this overall ocean floor mystery. Acceptance of any biblical theme, in my mind, should logically associate with the scientific evidence relating to it.

Of course, when I first discovered the face on the Pacific Ocean Floor in 1980, I naturally looked at the other oceans. At that time, I couldn't tell what was there, because I wasn't sure what I was looking for or even what criteria to use. I learned criteria as I went along. *I discovered the face on the Pacific Ocean Floor without a tracing.* That is, there was no tracing until later. Hundreds of hours were spent before I could be comfortable with that which I now know is obvious. These images relate thematically. The evidence is nothing less than overwhelming.

ALL THE OCEANS IN THE WORLD

Above is our own unique connection of six (6) Rand McNally© ocean floor maps detailing all of the ocean mountain ranges on Earth. They are:

1. Pacific Ocean Floor
2. Arctic Ocean Floor
3. North Atlantic Ocean Floor
4. South Atlantic Ocean Floor
5. South Polar Ocean Floor
6. Indian Ocean Floor

****

Below each set of maps are the corresponding border tracings comprising four (4) sets of related images. They are:

- ❶ The face of a weeping man on the Pacific Ocean Floor
- ❷ A dragon blowing fire into his face
- ❸ The Devil attacking him at his throat
- ❹ A cherub with a flaming sword

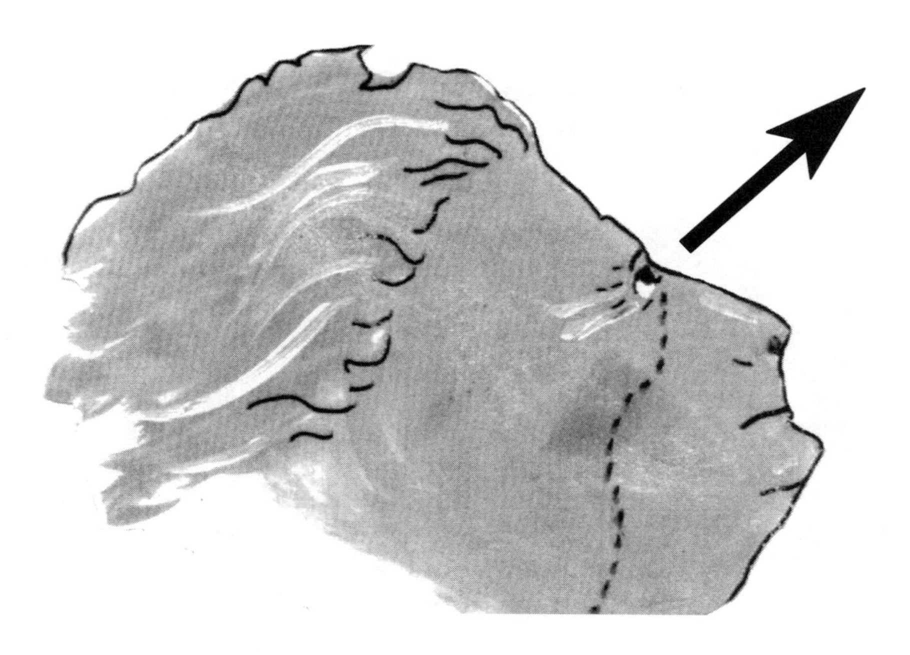

ARE THERE IMAGES ON THE LAND?

Throughout this research, I have continued to look for possible images on the land mass above sea level. Every continent has major mountain ranges and are undoubtedly full of clues. Rather than search aimlessly, I established some simple criteria and guidelines for this area of study. Concentration here would also be on border tracings and the deepest and highest elevations.

There has always been one major difference in my attempt to decipher the images I found on dry land. While studying the continents, I thought I knew what I was looking for. But when I discovered the face on the Pacific Ocean Floor, or the Devil in the South Atlantic, or the dragon, or the angel with a flaming sword, I didn't know what any of these four images might be before I started looking. Each one was a surprise.

When it came to the continents, I wasn't sure what the image of greatest importance might be, but I did know one thing about it. Whatever it was, it would be the *most important thing in the world.* My reason for knowing this was because of what I had already discovered with this research. I knew that the most stupendous phenomenon on this planet was the face on the Pacific Ocean Floor. Covering almost one-half of our world, this weeping man must be looking at something. What could it be?

Chapter 13

WHAT IS THE FACE ON THE PACIFIC OCEAN FLOOR LOOKING AT?

(Why Is He Weeping?)

It is the one question that is continuously in the mind of anyone who cares about this research. What is the face on the Pacific Ocean Floor looking at and what is causing him to weep? Newspaper reports have speculated that the face is looking heavenward. What do you think? Early in my research, I knew that whatever the face is looking at, it must be something that is *specific* and *easily recognizable.*

Once known, the answer should explain the second major question. This relates to the overwhelming tear-track (actually the East Pacific Rise) going down the right side of his face. Why does the face on the Pacific Ocean Floor appear to be weeping? What is breaking his heart? Curiosity abounds as to what might be the single recognizable thing that causes him to weep. The question is so abstract that it seems impossible to answer. How are we supposed to approach such a question?

Whatever the face on the Pacific Ocean Floor is looking at, I knew that it must fit certain criteria. There are criteria in discerning the images on each ocean floor (see page 16). *In each case, the tracing involved the entire ocean being studied.* Tracing is limited to the continental borders and the highest or most prominent mountain ranges. These tracings serve only to enhance or point to images that are already there. I have always said that the images that are on these topographic maps stand alone even without the tracings. Most serious researchers will quickly set my tracings to the side. Then they can study the images of each map on their own merit.

LOOKING TO THE LAND

What is the face on the Pacific Ocean Floor looking at and what is causing him to weep? I naturally wondered, could the answer to my question be found through topographic maps of the continents? The identities of each ocean floor are clearly identifiable and together cover about 70% of our planet. Only about 30% of the earth is left and it is all above the oceans and *on the land.*

The basic theme, found by examining topographic maps is: *Giant pictures, etched on the surface of the earth, substantiate the Bible story.* This theme compares favorably with the continuing story of "good versus evil" found throughout the Bible. First, the goal is to carefully study the geographic and topographic maps. The scriptures are then used as backup to help any preliminary suppositions make sense. I would like to say that this process has always gone smoothly, but it hasn't. What this study has always been about is a search for identifiable images that are already there.

The Pasadena Star News wrote that the face on the Pacific Ocean Floor is looking heavenward. This made sense at the time, so I looked toward the astronomical heavens. But I wondered, "why would the *most massive image on Earth,* the face on the Pacific Ocean Floor, weep because he is looking toward the heavens?" I felt that whatever he is looking at, must relate to the theme and criteria already established. I decided to start my search with the 33% of the Earth that was left: All land above sea level.

TRACKING HIS SIGHT-LINE WITH A GLOBE, YARN AND CELLOPHANE TAPE

Keeping it simple, I began by limiting my tools to three basic things: A globe, a one-foot length of black knitting yarn and some cellophane tape. I started at the place his sight line begins and then, not unlike a surveyor, began my search. Using a very small amount of tape, I affixed one end of the black yarn at the location of his eye which is near Manzanillo, Mexico. This kept most of the yarn loose so that I could freely shift position and experiment with the positioning required to logically track his sight-line.

THE LINE OF SIGHT NEVER ABANDONS THE SURFACE OF EARTH

Examining the surface of the globe with various topographic maps for comparison, I eventually would end my search. I resolved early not to experiment with any irregular lines or parabolic curves. I remembered that the gaze of the weeping eye of the face on the Pacific Ocean Floor is intense and focused. Whatever this face is looking at, *his gaze is straight and direct,* so I limited my search, using only straight lines. On a globe, these straight lines would follow the natural curvature of the Earth.

FINDING HIS LINE OF SIGHT

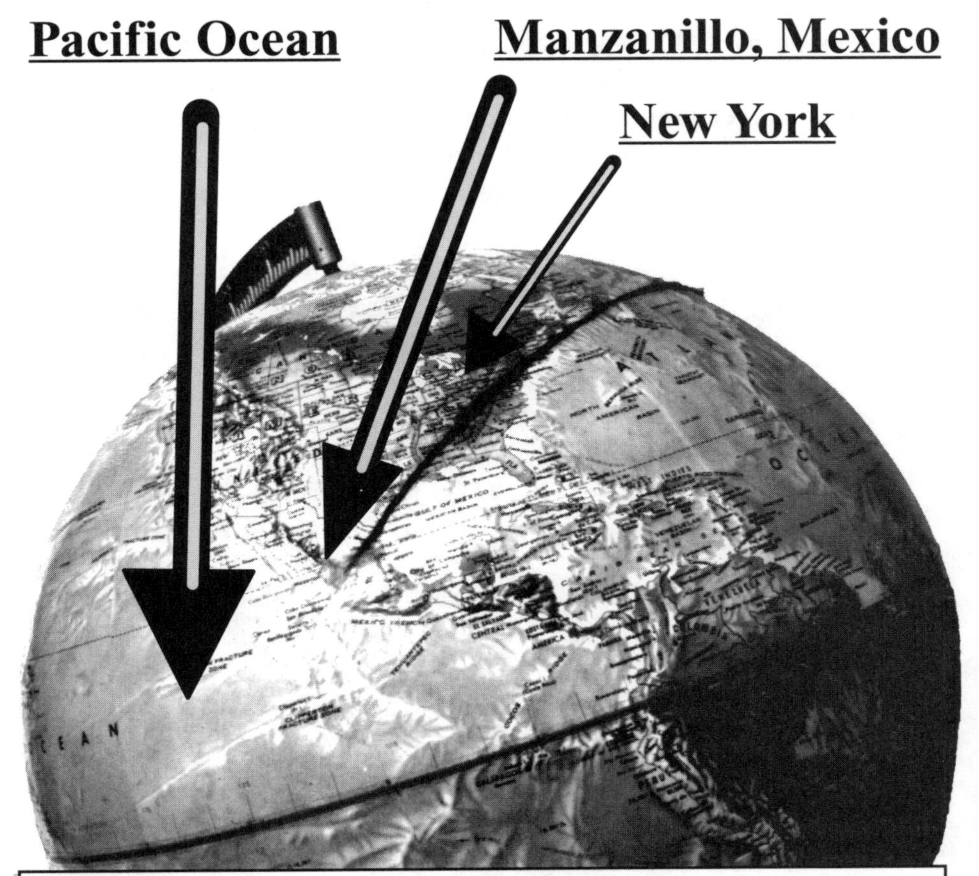

Black yarn on a globe was used to assist in tracking the direction of the sight-line of the face on the Pacific Ocean Floor.

After examining several alternatives, it soon became obvious that the face on the Pacific Ocean Floor is looking up *through North America, passing through New York State and then to some unknown point.*

If using ordinary office supplies to aid in my research seems simplistic, that is because it fits the nature of the work. Almost everything about this discovery is simple. It should be. The less complicated the theme, the more it can be appreciated by a wide and diverse audience.

33° BY 33°

My straight-line path kept aligning to the Middle East. The end of the path resulted in an arc of *about* 33° distancing *about* 33% of the Earth and ending at *about* 33° longitude and *about* 33° latitude. This point is just off the coast of Israel. In a later chapter, we will learn more about the number three and how it relates to this research. But, what is important now is not its specific location, but *about* where it may be found.

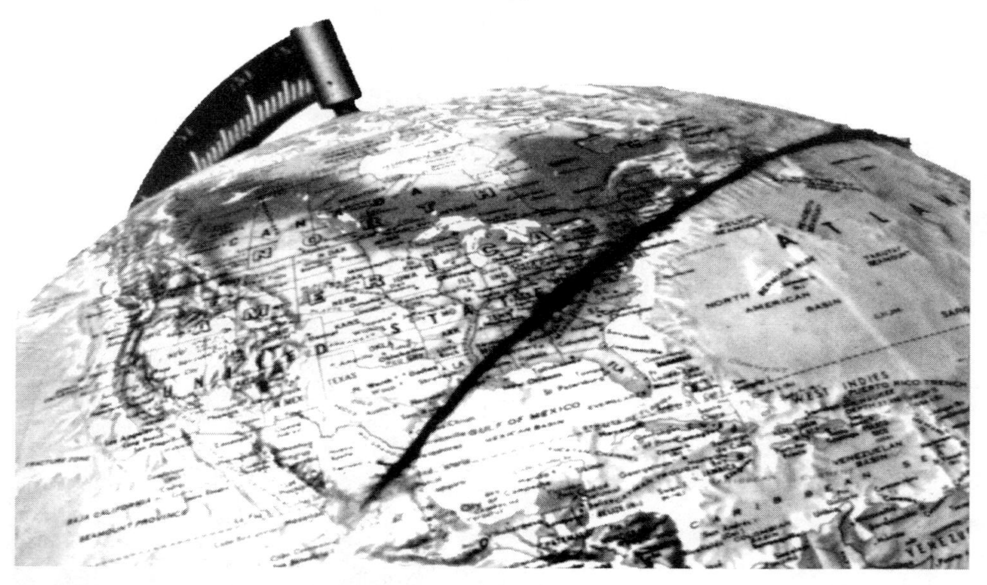

At first I asked myself why I should stop this particular straight line *about* 33 miles off the coast of Israel? Why not just keep going? I studied every possible ending point along this line and the 33° intersection made the most sense. Soon we will see why. As a Christian, the fact that this intersection is just off the coast of Israel (the birthplace of Jesus Christ) was particularly meaningful to me.

ABOUT, CLOSE OR NEAR

I acknowledge the importance of terms like, "*about*" or "*near*." I am not using these numbers to prove a point but rather to tell a story. Some generalities are acceptable when linking the exact nature of science to that of philosophy, which is fluid. At first, I didn't appreciate the significance of these numbers. I was hunting for any possible clue.

The 33° by 33° coincidence by itself would be insignificant if it wasn't for what I had already experienced. I kept remembering the (777) coincidence located at the mouth of the weeping face on the Pacific Ocean Floor. Also I thought of the (666) coincidence located at the mouth of the evil Devil on the South Atlantic Ocean Floor.

Since the path of his gaze was so direct and specific and also near Israel, my first conclusion was, "The weeping face is looking to Israel." If the face on the Pacific Ocean Floor is symbolic of God weeping, then God weeping for Israel made all the sense in the world. I found many scriptures that tell about God weeping for his chosen people. I kept imagining that the face weeps for Israel the same way God wept for the ancient Hebrews. Still, all the pieces did not fit.

For one thing, I couldn't find the *highly recognizable* shape of land mass I was looking for. Israel has no shape like anything immediately identifiable. Even if God looks to Israel and the Jews and Armageddon and Golgotha, none of these places fit my existing criteria. The answer *had* to have the detail and impact of the face on the Pacific Ocean Floor or I wouldn't be satisfied. So I set the globe, with the string taped to it, on my desk. Beside it, I placed an atlas opened to the Middle East for inspiration. I hoped that the answer might somehow come to me.

THE PYRAMIDS OF EGYPT

At one point, I decided that maybe the weeping face on the Pacific Ocean Floor might be looking at the pyramids of Giza in Egypt. I thought that this might be plausible for several reasons. First, the 33° arc created by the tape and yarn does pass through the Giza plane, which is at 30° longitude and 30° latitude. This is also the sight of the pyramids and the location that scientists consider to be the *center of all the land mass* on Planet Earth. Since the Great Pyramid is such a mystery to so many people, this further added to my fascination. What if somehow the mystery of the pyramids could be solved at the same time as the mystery of the face on the Pacific Ocean Floor?

After much consideration, I realized that neither Israel nor Egypt nor the pyramids could be the focus of attention of the face on the Pacific Ocean Floor. Although these places were important to the study, they did not hold the fundamental answers. Maybe later, experts in pyramid research may find more relevant evidence, but for now it is peripheral.

The sight-line of the face on the Pacific Ocean Floor passes through the pyramids of Giza, Egypt which are located near 30° longitude and 30° latitude.

SCIENTIFIC IN SCOPE

The Bible helped me qualify and add validity to the identities of the four main images on the our world's ocean floor. Modern science was necessary in my hunt for the geographic location of these images. To accomplish my goal, I needed topographic maps which were created by earth scientists. Still, I have always liked the way these four images, which cover about 70% of the earth, are recognizable to children with little knowledge of the Bible or science. The real beauty of this discovery is that although easy to understand, it is intrinsically based on scientific evidence.

The weeping man in the Pacific, the demon in the South Atlantic, the dragon in the North Atlantic and the angel in the Indian Ocean and Polar Seas connect geographically and thematically. Together these images reveal a story that is simple enough for a child to understand. The fact that the Bible gives validity to the story etched on the topography of our planet, calls for serious inquiry from adults. I realized that I must concentrate on the geographic and topographic maps, as well as satellite images of the Earth, to come up with the answers I was looking for.

These Four Images and the 70% of the Earth they Represent, Link Logically to the Answer of This:

Ocean Floor Mystery

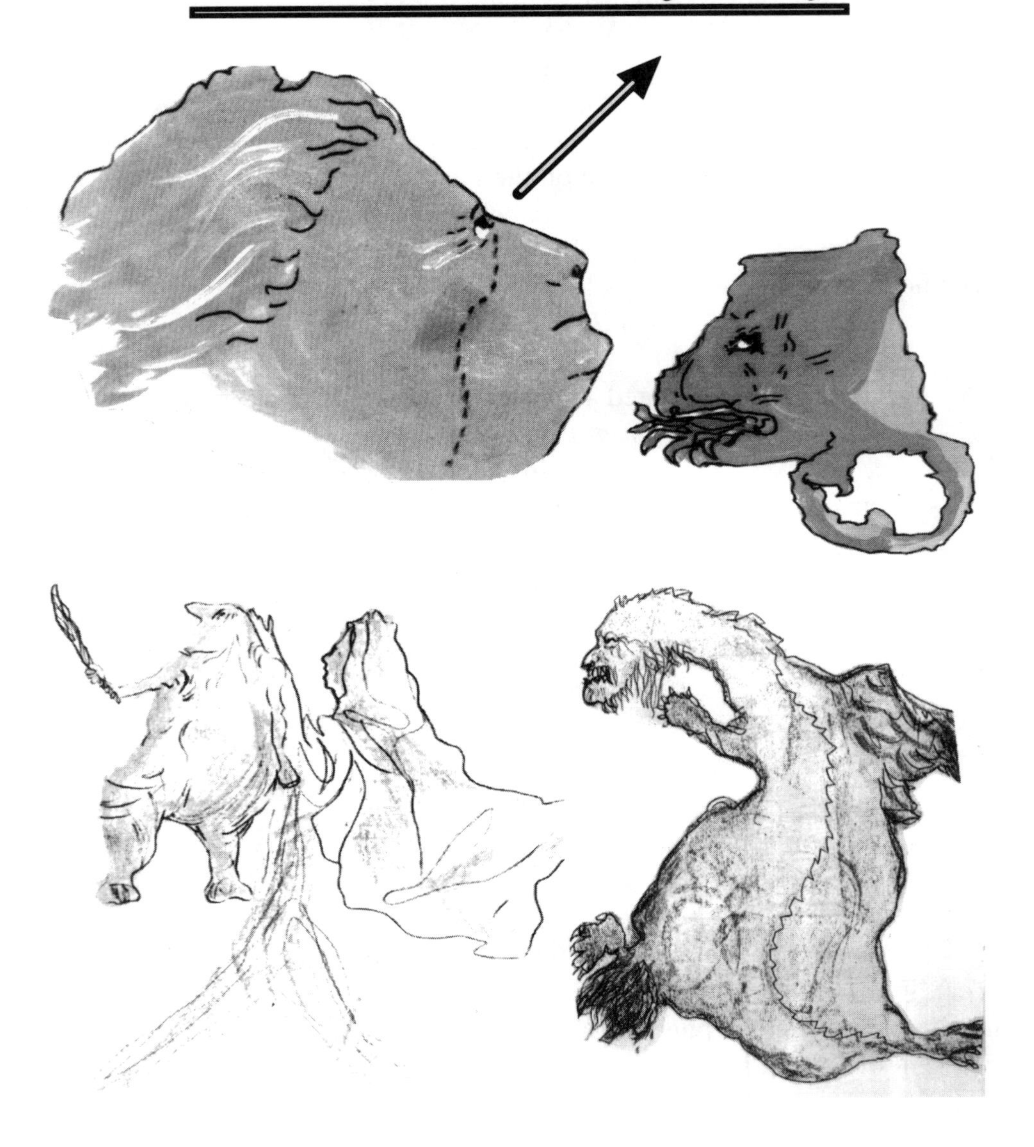

We are now ready to arrive at the conclusions we are looking for. That is, we must understand what it is that causes the great face on the Pacific Ocean Floor to be weeping. To do this, we shall contemplate seven remarkable phenomena.

Preview of New Facts

7 POINTS TO PONDER

Point #1

The topographical, weeping face on the Pacific Ocean Floor is looking at: *a young king, kneeling in prayer.*

Point #2

The heart of this young king which is at 33° longitude and 33° latitude is near *Jerusalem, Israel.*

Point #3

At his praying hands, located at about 30° longitude and 30° latitude, is the *Great Pyramid of Egypt.*

Point #4

The outline of this obedient monarch includes the entire Middle East and the *center of all the land mass on planet Earth.*

Point #5

A careful study of this king's face shows that he appears to have the exact *likeness of a lamb.*

Point #6

The Suez Canal is the only man-made feature of this research: a tracing of *a nail at his wrist.*

Point #7

Through careful study, we find that the identity of the young king in prayer is: *self evident.*

Who cannot see that there really is a weeping face on the Pacific Ocean Floor? Even if one is not a believer in demons, it is hard to dispute that the arrangement or phenomenon on the South Atlantic Ocean Floor does represent the Devil. Also, it does appear to be a dragon etched on the North Atlantic and the Arctic Ocean Floors. And even if the giant angel with the flaming sword on the Indian Ocean Floor is not as recognizable as the other ocean floor images, it is interesting to study and think about. What can we think of about these images? Four big pictures logically link and encompass 70% of our world. If you are challenged by these four images, what is ahead will not be a disappointment. This is because, even though we have discovered so many new things, there are also many unanswered clues that will help bring to a close, one of the greatest mysteries of all time: *What is the humongous face on the Pacific Ocean Floor looking at and why does it cause him to weep?*

Ahead, we will see how the involvement of the pyramids, Israel, Egypt and the entire Middle East contribute to the significance of the image on the Pacific Ocean Floor. Now some of the previously mysterious longitudinal and latitudinal lines will become more understandable. Additionally, we will see how the Holy Bible helps to identify this image, sometimes in ways not expected.

FOCUSING ON THE MIDDLE EAST

On the next few pages we will see first a full world view flat map. Then, using the same map, we will come in closer to the area of the Middle East. What we are about to look at is not a topographic map. What is pleasing about this map is the clarity of the continental and national outlines. For the most part, countries are not identified. This is perfect for our purposes, because at this stage, our main area of concern are the shapes of the land masses and what they might mean. Later we will look more closely at the actual land formations on a topographic map. Although identification of all of the continents is a worthy goal, for now we will focus on the area of the entire Middle East.

The style of this map is also called a Mercator Projection. It is how map-makers show our spherical world on a flat map. That is why the areas most accurate in relative size are near the middle latitudes. Things become more distorted at the edges and toward the poles. That is also why the overlay of the weeping face doesn't fit perfectly. If we were to be completely accurate in every way here, the weeping face would be looking straight horizontally, even though on a globe, he is looking up and over to the Middle East. Still, by showing placement, we can get a pretty good picture of the theme.

Highlighting the Middle East

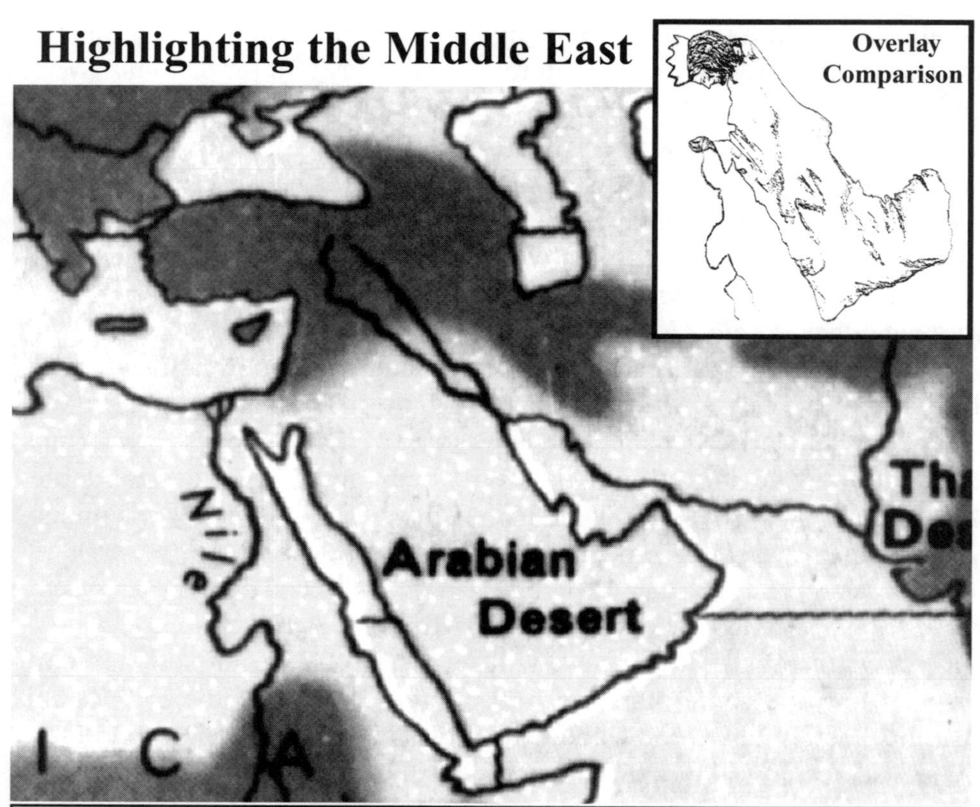

This is from the same Rand McNally© map as shown on the next page. It is cropped to the area that is in the black and white square. Above right here is our artistic overlay of a king, kneeling in prayer, with blood dripping through his nail scarred hands.

THE DAY I FOUND THE ANSWER?

One day as I passed my desk, I kept glancing at my fully opened atlas depicting the Middle East. I thought, "Whatever the biggest figure on this planet is looking at, it must be the most important thing in the world." Was my thinking too small? I guessed that the weeping face could not be looking at any country or nation. *But what if the focus is on a shape?* I backed away from the atlas, still looking at it. Then I thought, "Look at all of it." Then, without warning and with the same clarity as when I first discovered the face on the Pacific Ocean Floor, I saw the image that solved the riddle. I saw the young king kneeling, head bowed and praying.

Soon, I resolved to decipher this image and declare its identity? I found that the relating factors that correspond to the identity of this figure are overwhelming. Many of these factors will be expanded upon in the next chapter. It was for these reasons and more, which convinced me of one thing, the weeping face on the Pacific Ocean Floor is looking at:

A Young King Kneeling in Prayer

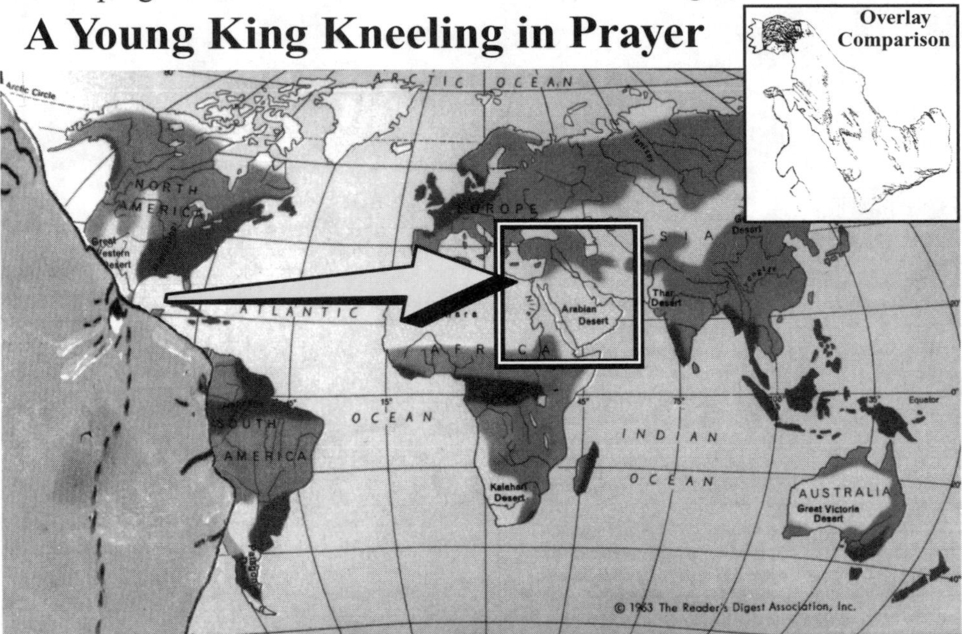

Here we inset our overlay of the weeping face on the Pacific Ocean Floor. In the black and white square is the general area of the Middle East that is enlarged on the previous page. At the top right of this page is an artistically enhanced border tracing showing the crowned king, kneeling in prayer. The map on this page is a Rand McNally© world moisture map, which emphasizes the continents. The shading represents levels of land rainfall. On land, the lighter areas have the least moisture and the darker areas, the most.

The entire Middle East and its adjacent areas include sections of northwestern Africa, the western Mediterranean and some of the west Asian and Eastern Block states. Once we see that the shape of the entire Middle East is of a young king in prayer, it is then possible to identify him. Study the shape and notice the comparisons. All of the images in this chapter relate to the shape of the Middle East. Flip back and forth between pages. What else do you thinkt the center of all our planet's land mass might be?

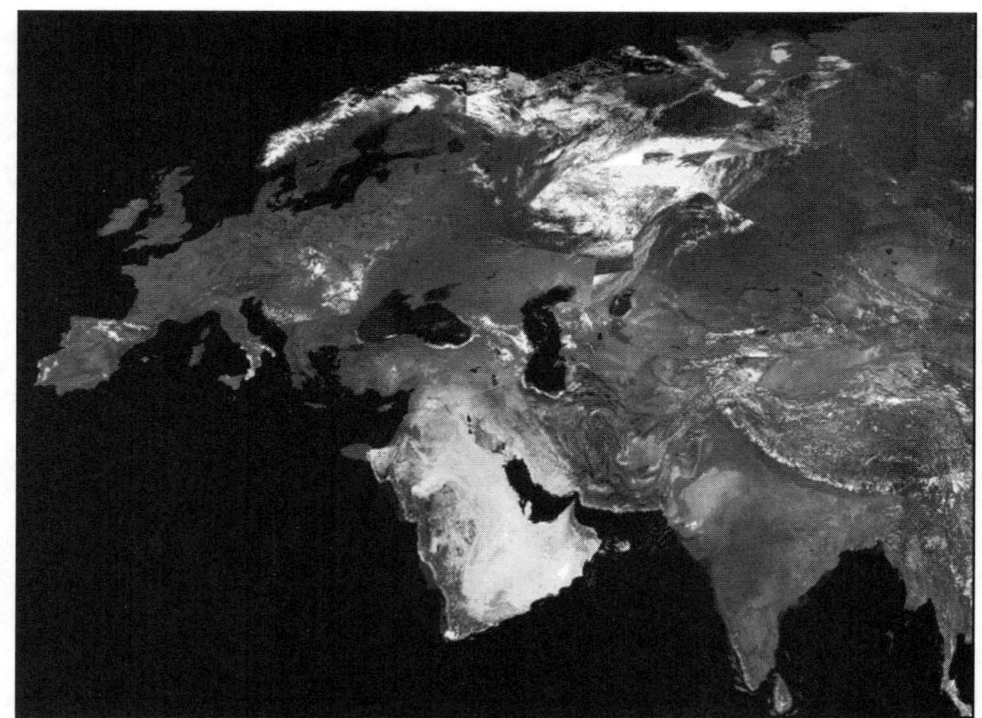

This is a satellite view from space showing the topography of the entire Middle East, with Europe above it and Asia and Russia to its right. The young king that is kneeling obediently in solemn prayer is clearly visible. U.S. Defense Mapping Agency, NASA.

Guessing in this case would not do. I decided that before I could even speculate on the identity of this obediently solemn figure, more evidence would be required. Topographic maps would have to be studied in detail. For weeks I became almost obsessed in this area of the research. I found myself locked in focus on the young king kneeling in prayer. He is a *young* king because as we will soon see, the evidence suggests not only his identity but also his age. We will also learn why this matters. Imagining the identity of this kneeling king is not difficult. A general knowledge of the Bible can be very helpful here. After that, all it takes is a closer look at various maps of the Middle East.We can see that the outline of the Middle East is clear, but more importantly, this image fits our basic outline criteria and fits our overall biblical theme like a glove.

Chapter 14

THE OBJECT OF HIS AFFECTION

In the previous chapter, we were comparing the *outline* of the Middle East with our artistic overlay. In this chapter, we will study various views of the topography of the Middle East. Also, we are going to look at the same area identifying some of the nations and countries that are there.

The Middle East

An Outline Map

This is a cropped enlargement of the Rand McNally© map (from page 164) highlighting on the Middle East. In the upper right corner is the artistically enhanced border tracing. This area (also shown on the next page topographically) is the geographical center of all the land mass on Earth.

The Middle East

A Topographic Map

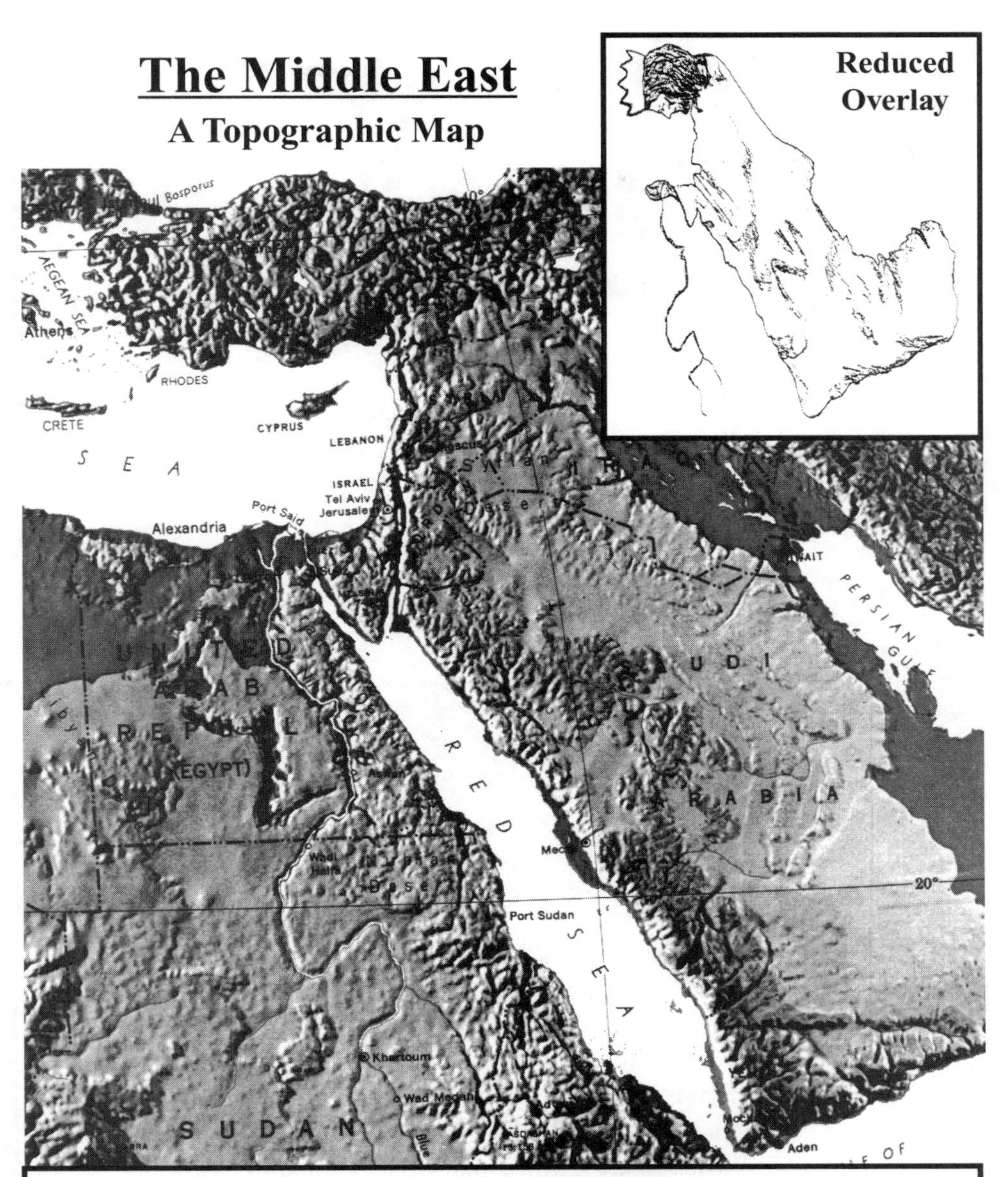

This is a Rand McNally© map of the Middle East area. It shows the mountain ranges of the entire region, without bodies of water. On the next page, we will examine the upper torso section of this young king in prayer and we will also use this map.

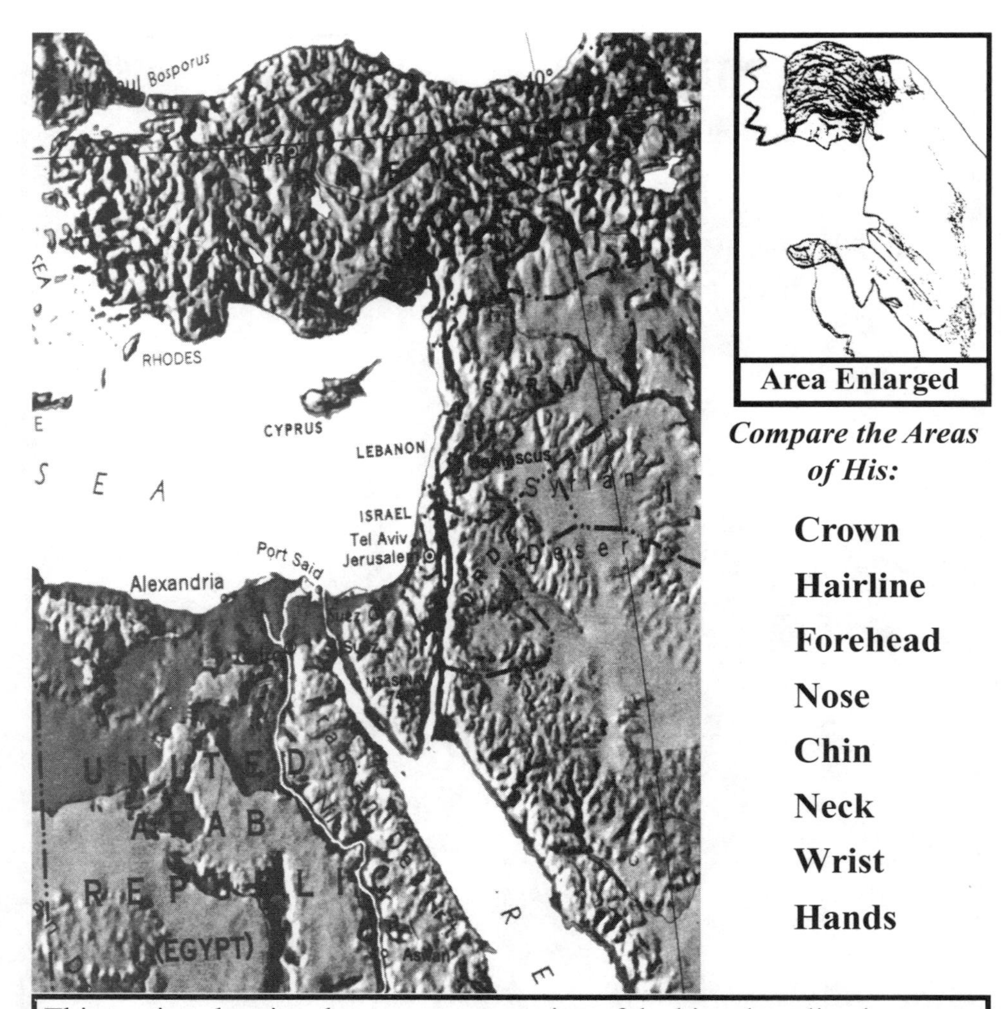

This section showing the upper torso region of the king, kneeling in prayer, provides a closer look at the Rand McNally© map of the Middle East. Here, we see that his crown is at northeastern Turkey, his heart is near Israel and his praying hands include all of Egypt. His wrist is at the Suez Canal and the sleeve of his garment includes all of the Sinai peninsula. Although we do not see clear facial features on this map, there are still enough recognizable points to help with our identification. Take some time studying this image. Consider that the sight line of the Pacific points directly to the kneeling king's heart, as shown in the last chapter. It does provoke contemplation.

Where Are His
Eye And Mouth?

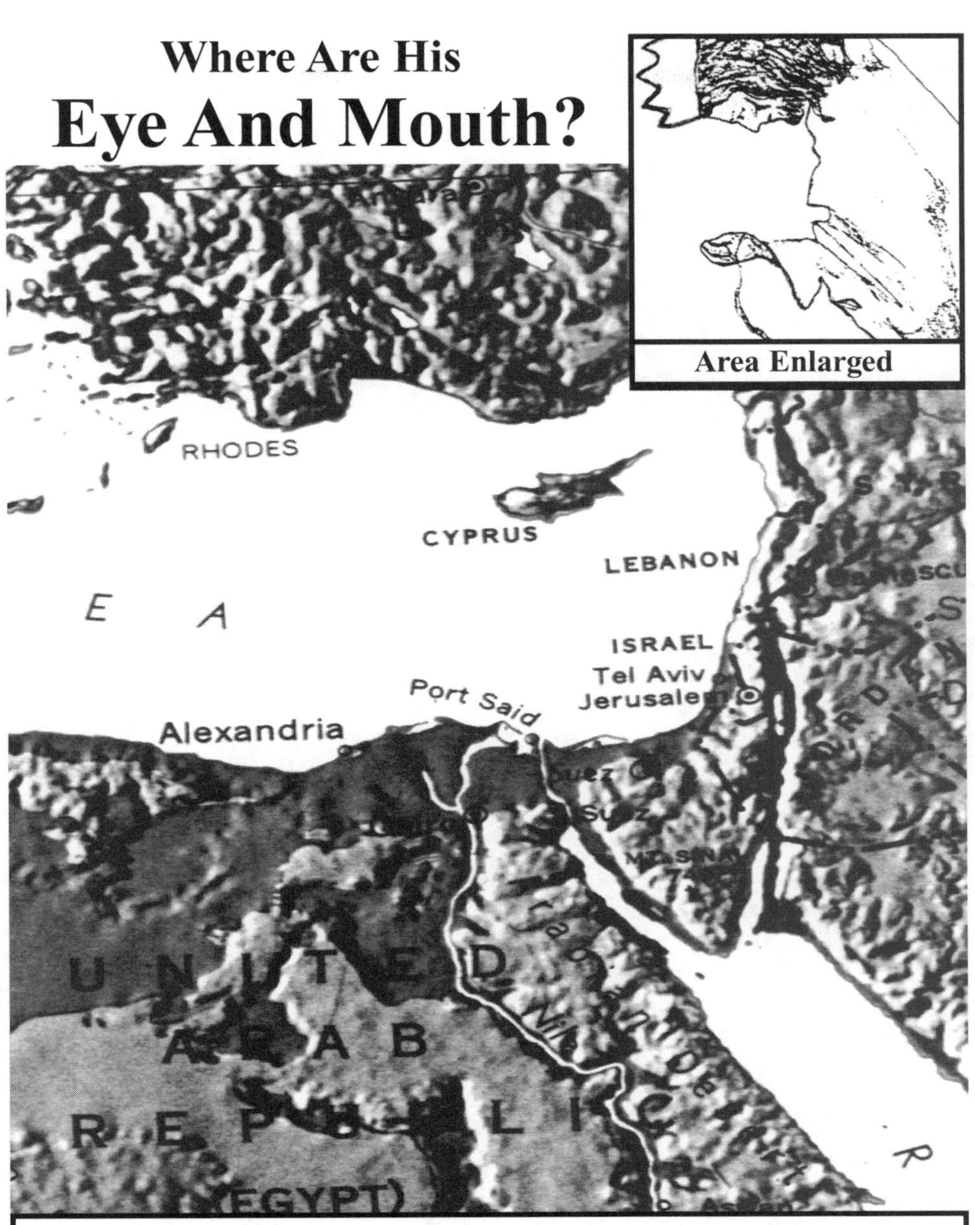

The reason we do not see his eye or mouth on this map is because it shows only the mountains and less detail of the rivers or lakes. On the next few pages we will study his face with a different map.

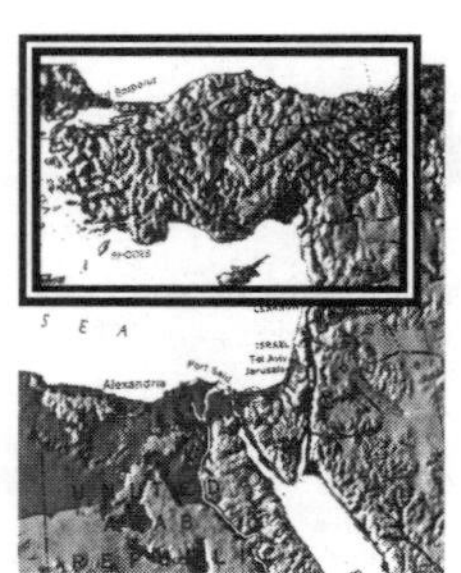

This topographic map was made with computer software and mapping information from the Department of Defense, the Central Intelligence Agency, NASA and other sources. It shows the

Head and Face

of the young king in prayer.

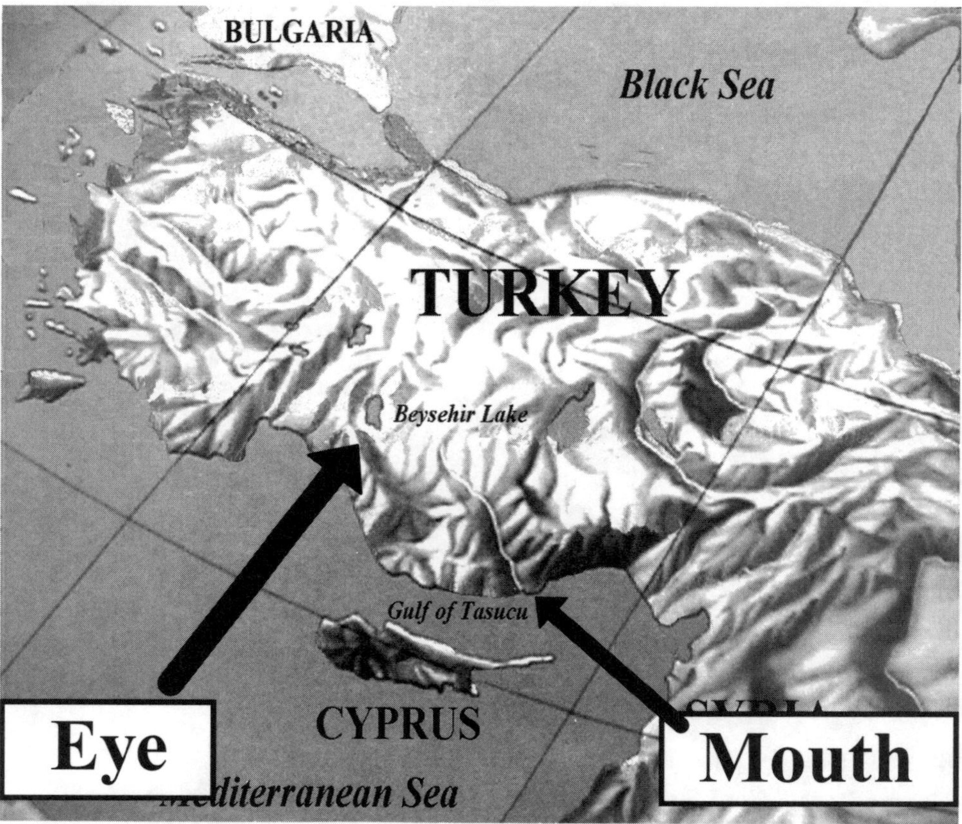

At the top-left of this page is the topographic map from the previous page. At the top-right is the overlay of the king. In the black and white square of each is the area magnified at the middle of this page.

On the next page, we magnify the areas of the

Eye and Mouth.

Here below is a magnification of the "face section" of the map shown on the previous page. Study it for a moment. It is highlighting the area directly above Cypress. The comparison is shown in the black and white box located here.

His Eye is at Beysehir Lake.
His Mouth is near the Gulf of Tasucu.

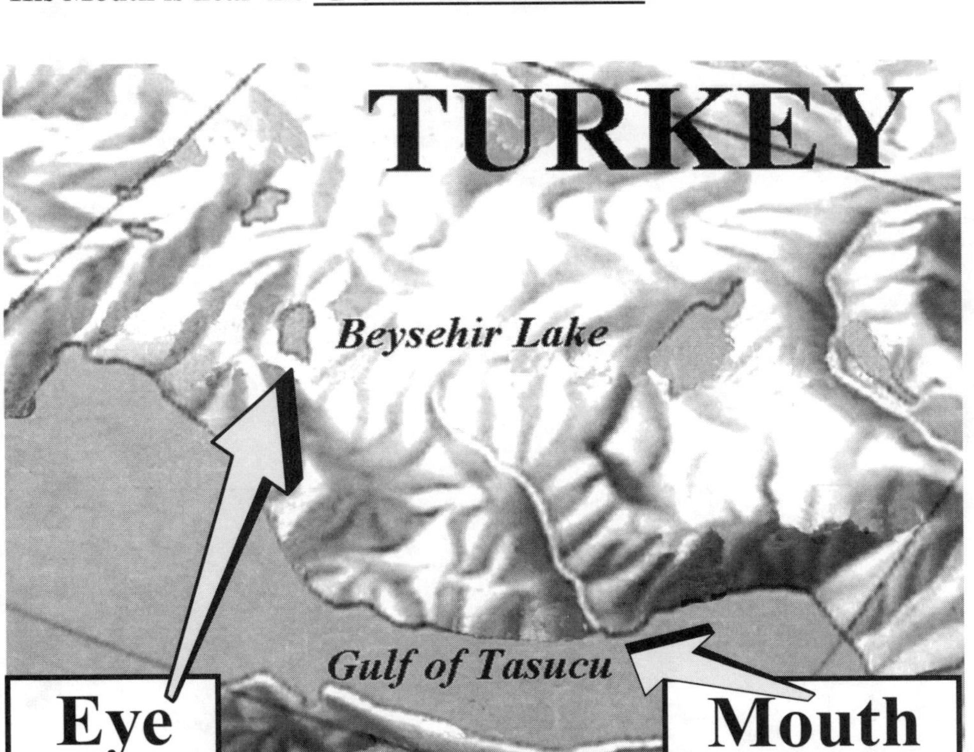

This map does an excellent job of showing his **mouth-line**. The crevice is separated by mountains. Here also is the outline of his **eye** at Beysehir Lake. Can you see his **eyebrow** and **forehead?** Notice that the topographic face on this map looks more animal-like than human-like?

Could This Represent a Lamb?

This is a cropped section of the face area of the map on page 172, with the overlay next to it. Below is an extreme close-up of

Beysehir Lake which is his Eye.

Eye

What King do you know of that is also called:

The Lamb?

??????????????

Chapter 15

HIS PRAYING HANDS
(The importance of a nail)

Near the point where the blood drips through his nail scarred hands is the Great Pyramid of Egypt. Many pyramid enthusiasts may get very excited about this coincidence, but what really helps identify this young king in prayer is not his relationship to the Great Pyramid of Egypt, but rather the nail that is going through his wrist at the Suez Canal. The purpose of this chapter is to elaborate on that idea.

On the next page are two maps by Piazzi Smyth which show that the Great Pyramid is located at the geographical center of the land surface of the whole world. The map on the bottom shows that this is at the intersection of 30° longitude and 30° latitude. The evidence shows that this is also the same location as the tip of the praying hands of the young king kneeling and praying. This added to the fact that at the heart of the worshipping king is 33° longitude and 33° latitude, help even more in clarifying his identity. As we will see by the end of this chapter, there are many interesting additional pieces of evidence that relate to the topography of the Middle East, which is also this young king with his head bowed in solemn prayer.

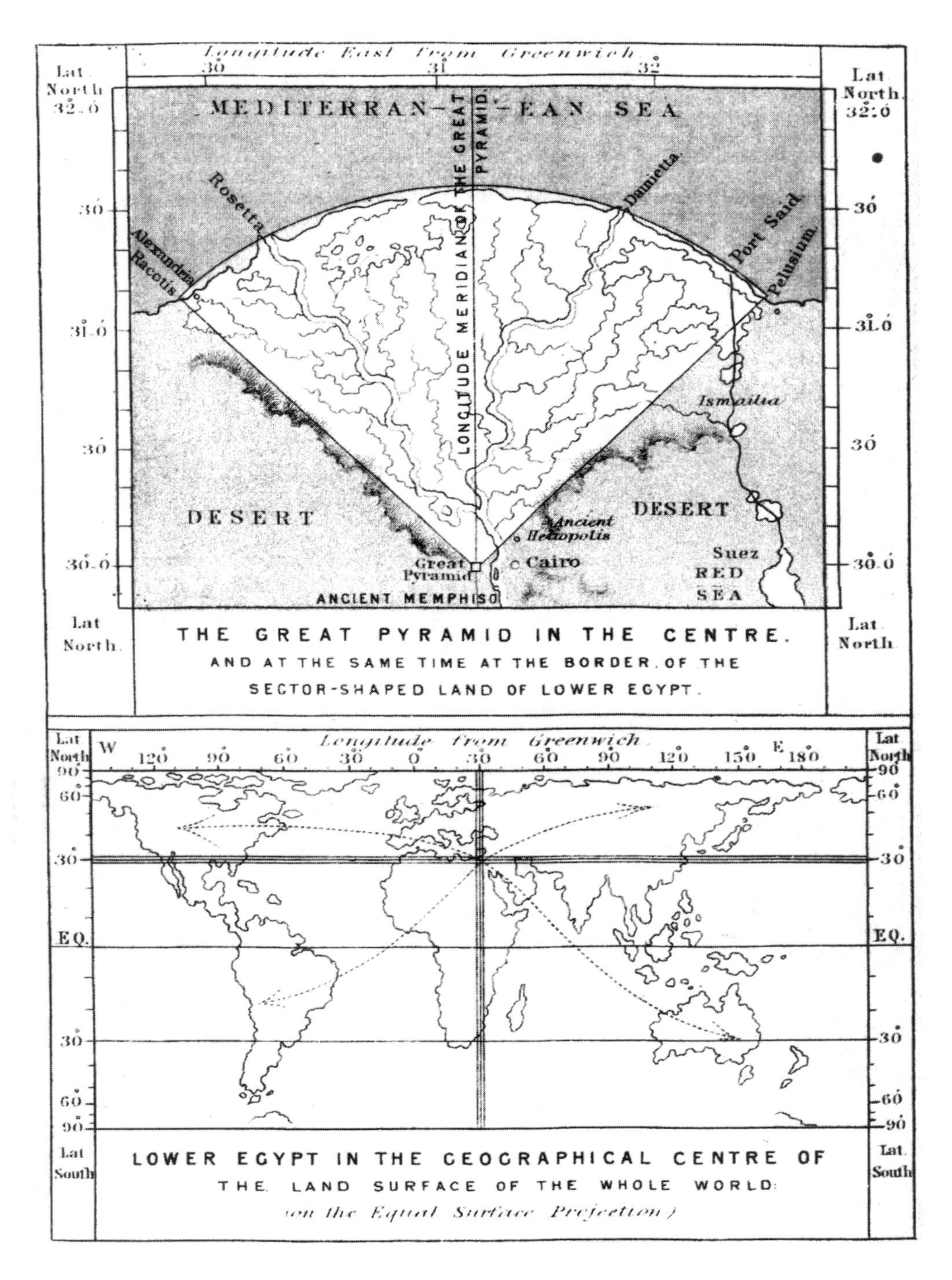

THE GREAT PYRAMID IN THE CENTRE.
AND AT THE SAME TIME AT THE BORDER, OF THE
SECTOR-SHAPED LAND OF LOWER EGYPT.

LOWER EGYPT IN THE GEOGRAPHICAL CENTRE OF
THE LAND SURFACE OF THE WHOLE WORLD:
(on the Equal Surface Projection)

Compare the elements on this page with those on the next page. The map below is an enlargement of the praying hands section there. The area of emphasis is known as Port Said at the Suez Canal in northern

EGYPT

What follows is more evidence that this might be his

Praying Hands

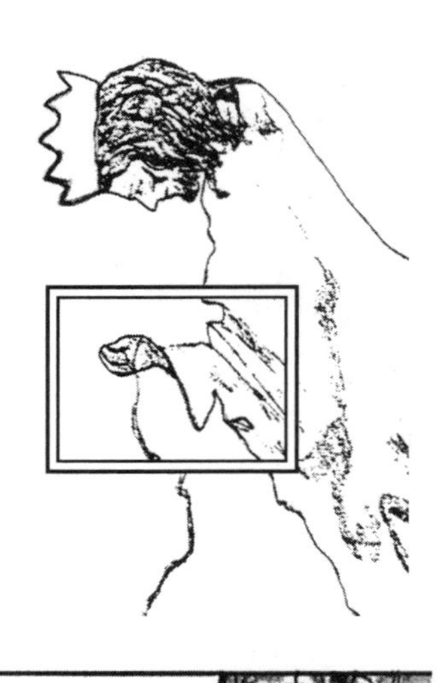

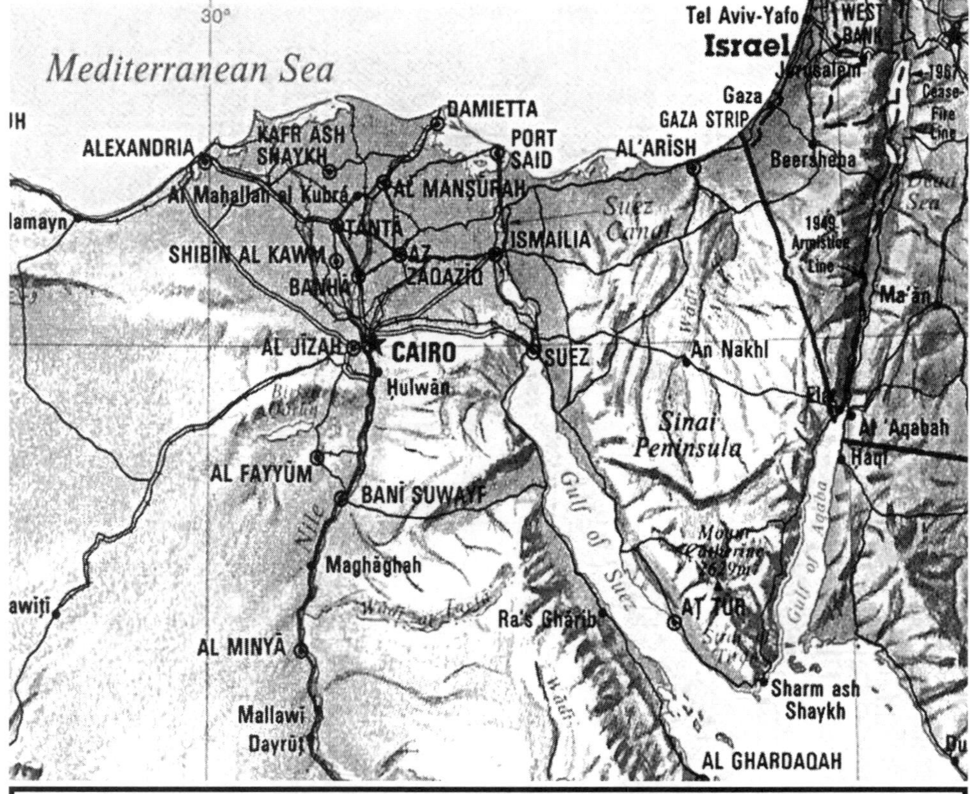

The maps on this page and the next are courtesy of the U.S. Central Intelligence Agency.

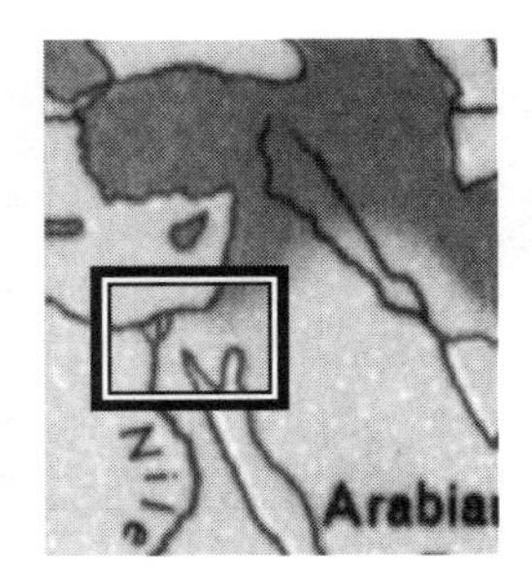

A COMPARISON

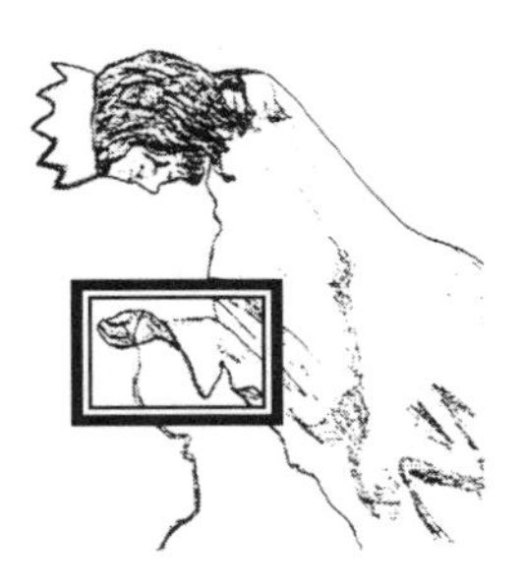

At the left is a section of that same Middle East map as the one shown on page 168. At the right is a sectional artistic overlay tracing. Together, they help as references when examining the more detailed map below.

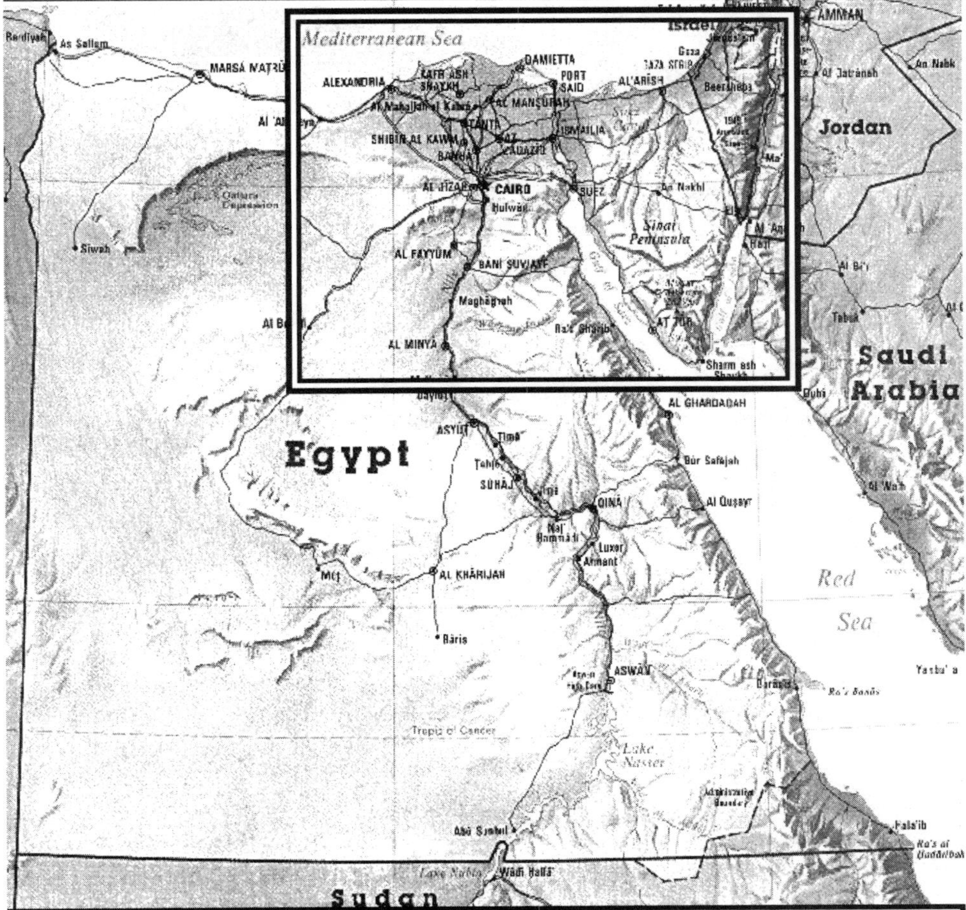

This is a Central Intelligence Agency map of Egypt detailing cities, waterways and topography. The area in the black and white inset square above highlights the praying hands of the young king who is bowing down. To view it close up, see the next page.

The map below is the same sectional map seen on page 179. On the next page is a praying hands overlay created to emphasize the

Suez Canal

as the schematic of a wound that might come from a

<u>Nail</u>

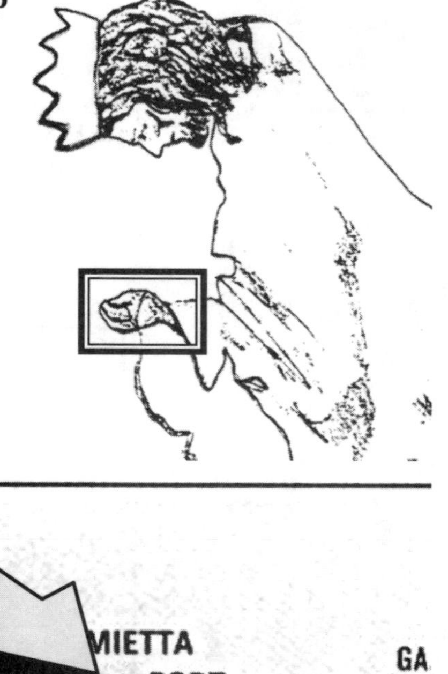

The only man-made part of this entire maps research is the Suez Canal. It reveals an uninterrupted wound at his wrist, which drips at its bottom into the Gulf of Suez and also into the longest river on Earth, the Nile. On the next page is an overlay showing an artistic tracing of his praying hands. Above it is the Suez Canal area of the map on this page. Next to that is a comparison with the stylized nail from the overlay.

John 20:25
"...Except I shall see in his hands the print of the nails, and put my finger into the print of the nails and thrust my hand into his side, I will not believe."

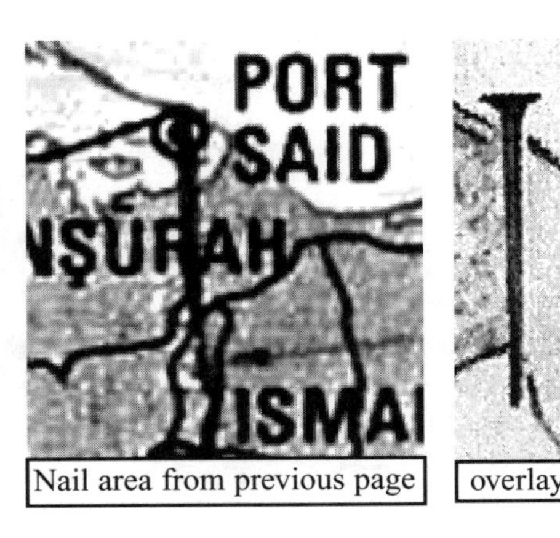

Nail area from previous page

overlay

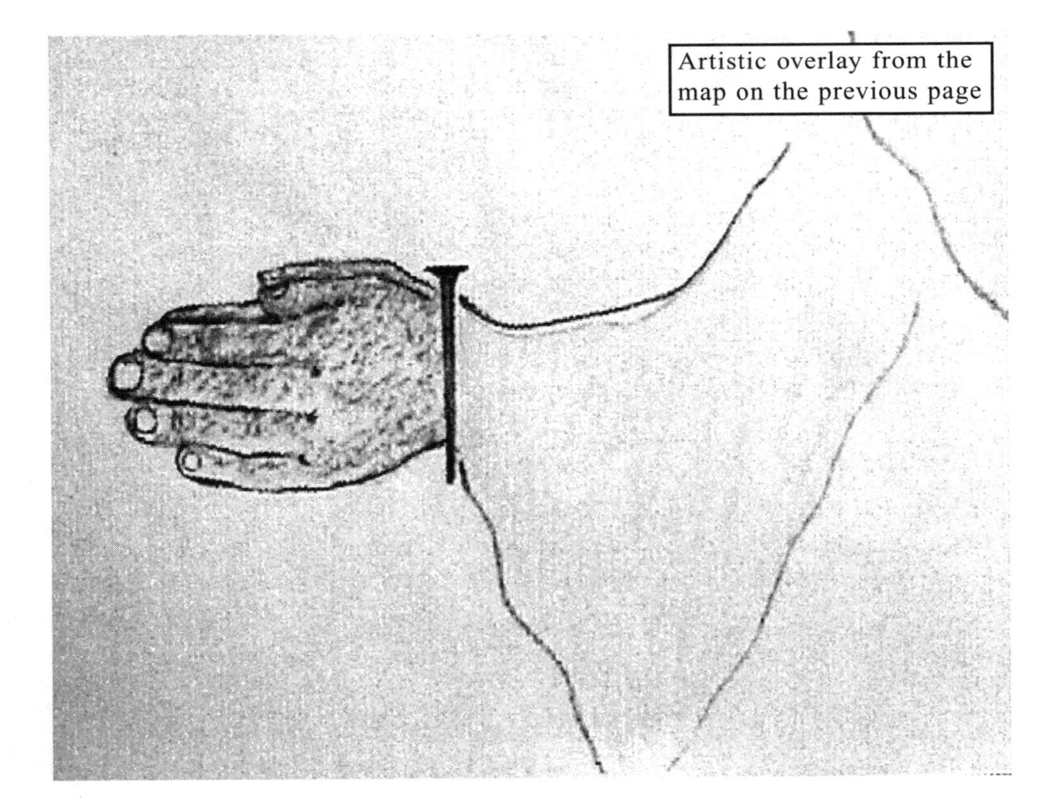
Artistic overlay from the map on the previous page

Chapter 16

JESUS AND THE NUMBER 3

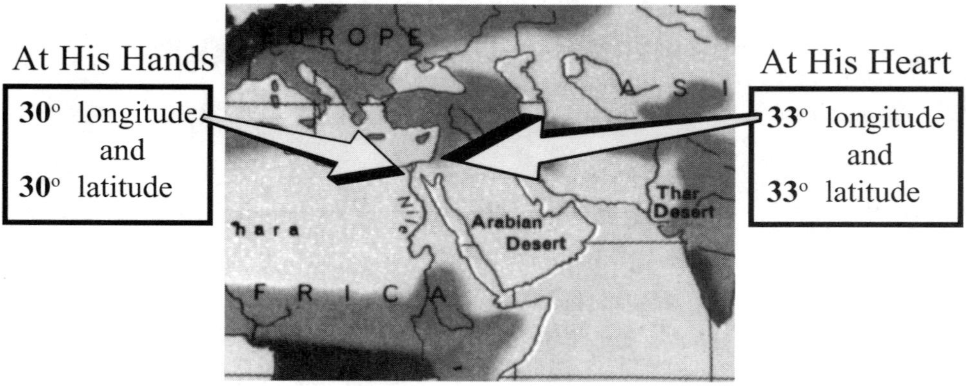

At His Hands

30° longitude
and
30° latitude

At His Heart

33° longitude
and
33° latitude

3333333333333333333

At this point in our study, it becomes necessary to begin our hypotheses on who this young king might be. Let us ask ourselves: What other young king do we know of who has a nail going through his wrist and who would kneel humbly in obedient prayer? What other king is also known as "The Lamb," who could also have his heart at Israel? What other clues can we find? I believe that the preponderance of evidence on the topographic maps of the Middle East indicates his identity. I hypothesize that the young king is most likely a symbolic picture from God, forged on our world, of his one and only begotten son: *Jesus Christ at the time of his crucifixion.*

I realized from the beginning that there is nothing on the surface about this kneeling king that identifies his youth. But the coincidences of pertinent longitudinal and latitudinal lines reminded me of the many coincidences that relate to Jesus Christ, his life and his plan.

The sight-line beginning at the eye of the face on the Pacific Ocean Floor leads to **33°** longitude and **33°** latitude, which is just off the coast of Israel. This **33°** by **33°** intersection points directly to the *heart* of the young king in prayer. We have also seen that at his praying hands is the intersection of **30**° longitude and **30**° latitude which is also the location of the Great Pyramid of Egypt.

If this image does depict Jesus Christ at the age of his crucifixion, then I knew the term "young king," made sense. The Bible reveals that Jesus was **33** years old at the time of his crucifixion and resurrection. The evidence of a nail at his wrist shows that this young king, on the topography of the Middle East was also crucified. But there is more about the number **3**, as it relates to Jesus Christ, that helps us identify the outline of the area encompassing the Middle East.

JESUS AND THE NUMBER 3 IN THE BIBLE

For many years I have been fascinated by the coincidences relating to the number **3** and Jesus Christ. According to the Bible, here are some of them:

1. Jesus was about **3** years old when his parents fled to Egypt.
2. At 12 years of age Jesus left his parents and they went back for him after **3** days.
3. Jesus was **30** years old when he began his ministry.
4. The ministry of Jesus lasted for **3** years.
5. He died on the cross at **33** years old.
6. He was ransomed for **30** pieces of silver.
7. He was on the cross for about **3** hours.
8. He gave up the ghost at about **3:30** in the afternoon.
9. He arose from the dead on the **3**rd day.
10. He brought back people from the dead **3** times.
11. Jesus asked Peter **3** times, "Do you love me?"
12. In the Gospels, Jesus was tempted by the Devil **3** times.
13. When Jesus Christ was on the cross, on that day alone, he fulfilled **33** prophecies.
14. Jesus knew **3** women by the name of Mary.
15. The Sun, which Jesus is compared to, is **330,330** times larger than the Earth.
16. When Peter denied Jesus, the cock crowed **3** times.
17. The word Christian appears in the Bible exactly **3** times.

MORE **3**'S RELATED TO JESUS AND HIS PLAN

1. Church history indicates that there were **3** wise men that visited the baby Jesus.
2. The word God is in the first chapter of Genesis (in the King James Bible) **33** times.
3. Only **3** angels are specifically named in the Holy Bible: Michael, Raphael and Gabriel.
4. On **3** occasions the people of Israel reaffirmed their covenant with God before Moses climbed Mt. Sinai.
5. God made **3** promises to Abraham: a great nation, family seed, and that he would be known all over the world.
6. It took **3** months for Moses and the Hebrews to reach Mt. Sinai after they left Egypt.
7. The Trinity of God the Father, Jesus Christ and the Holy Spirit are considered **3** persons with one nature.

****

Chapter 17

A FATHER AND SON RELATIONSHIP

The message that is revealed by looking at the images on our planet, through the study of topographic maps, supports the biblical theme. It is the story of a father and son relationship. The weeping Pacific, as the Father looks to the Middle East, which represents the obedient Son of God. It is like the final obedient prayer Jesus said to his father in the garden of Gethsemane on the night he was betrayed and captured. He knew he was to accept a painful death and be slain as a sacrifice for all of humanity. He did not want this, but he realized that it must be. The passage reads:

Matthew 26:39
"And he went a little farther, and fell on his face, and prayed, saying, Oh my Father, if it be possible, let this cup pass from me: nevertheless not as I will, but as thou will."

This is also a story of God's love for humanity through his only son. The Son of God became flesh and lived among humanity. The death and resurrection of Jesus Christ is central to a divine plan of salvation designed to guarantee eternal life to all repentant souls. According to the Bible, every human being is born into a sinful world and has a sinful nature. Christianity teaches that because of the original sin, we are already doomed to eternal damnation and an everlasting absence from God. To choose the world and its trappings, is to choose darkness rather than the light of God and eternal salvation from the sacrifice of his son, Jesus Christ.

John 3:18
"He that believeth on him is not condemned: but he that believeth not is condemned already, because he hath not believed in the name of the only begotten Son of God."

The Father and Son relationship depicted on the mountain ranges of our planet tell a story that matches the scriptures. Also matching the Bible story is the Devil on the South Atlantic Ocean Floor, the dragon on the North Atlantic Ocean Floor and one or possibly two angel type creatures located on the Indian Ocean Floor. But the focus of the Bible and the picture story chiseled on the topography of our planet is about a Father and Son and their roles in the ultimate sacrificial drama.

WHO IS GOD?

It can be overwhelming to see confirmation of a weeping Father and a nail scarred Son. Beyond that, we know that God is Spirit. Together these three identities make up what is known as the Trinity. No person but God knows everything about God, but because of the Bible, there is much that we do know. So, who is God? To explain what every religion knows about God is impossible. What the various Christian and non-Christian religions teach may be found in most public and university libraries. One thing is clear. God loves this world so much that he gave his only Son as a ransom.

John 3:16-17
"For God so loved the world, that he gave his only begotten Son,
that whosoever believeth in him should not perish,
but have everlasting life. v17 For God sent
not his Son into the world to condemn the
world; but that the world through
him might be saved."

I address some of the most fundamental tenets of the Christian faith throughout this book because I am a Christian. My research of natural land formations on the Earth convince me that the truth of Christianity and the Bible are reconfirmed. In this chapter we will address some of what the Bible says about God and in the next chapter, what it says about Jesus. Still, the readers are encouraged to research the deity independently. The study of God is a lifetime process. I believe that anyone with a proper heart can find God. I wish that I could somehow simplify the identity of God. Since that is impossible, here are some of God's most identifiable attributes.

For God So Loved The World That He Gave His Only Begotten Son...

A Father and Son Relationship

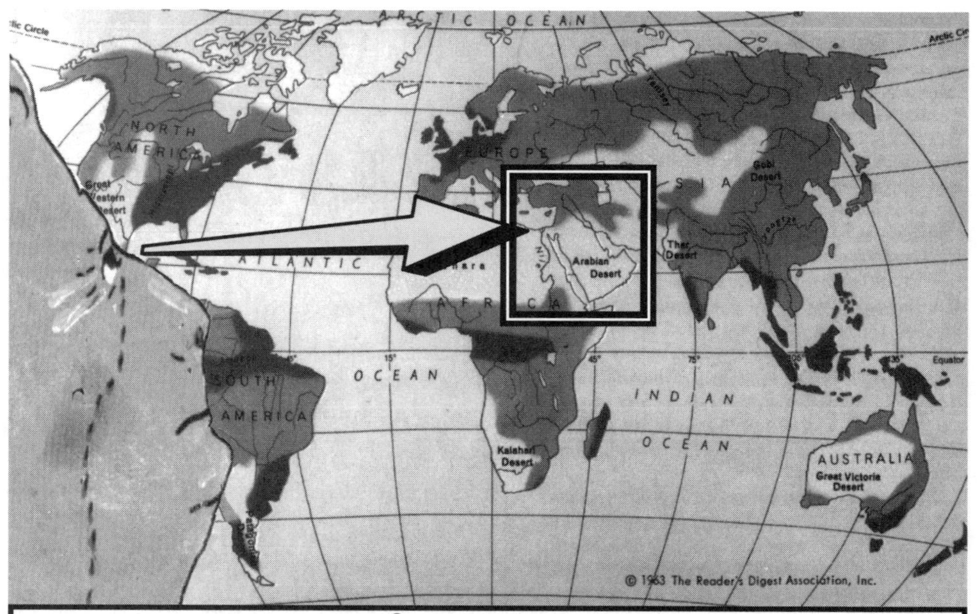

Above is the same Rand McNally© world moisture map that is shown on page 165. At the left of this map is the face of what could be construed as a symbolic image of God the Father. He weeps while looking to his only Son, represented by the figure in the black and white square at the Middle East. God looks to his son Jesus, who is bleeding at his wrist because of the nail wounds. Jesus has suffered the crucifixion for all the sins of humanity. At his praying hands are the pyramids of Egypt. His face has the likeness of a lamb. He is the obedient, worshipping king that has a nail going through his wrist at the Suez Canal. Careful study shows that the blood dripping through his wrist becomes the Gulf of Suez and the Red Sea and the longest river in the world, the River Nile.

GOD THE FATHER

As much as some people might wish differently, the Holy Bible also specifically calls God our Father. That is why Christians know it is appropriate to refer to God as "he." He is our heavenly Father. This was confirmed by Jesus (Matt. 11:25, Matt. 28:19, Mk. 14:36, Lk. 23:34, Lk. 23:46, Jn. 1:14, Acts 15:6, I Cor. 8:6).

GOD IS ETERNAL

The Bible explains that our Heavenly Father God is forever eternal (Gen. 21:33, Deut. 32:40, Deut. 33:27, Psa. 9:7). King David proclaimed this to all of the hosts of heavens in:

Psalms 90:2

"Before the mountains were brought forth, or ever thou hadst formed the earth and the world, even from everlasting to everlasting, thou art God."

In the Bible God is viewed, not as an abstract or impersonal principle, but as the living, active Lord of all human history. God's power over the universe is traceable to his being its creator and an integral part of all life. God the Creator, as seen in the Old Testament, becomes God the Redeemer in the New Testament.

We can never have perfect knowledge of God in this life, but we are able to know certain qualities and characteristics that are attributed to him. We can do this on the basis of our reflection of his revelation of himself in history and Holy Scripture. Whenever God is revealed to humans, it is often in a unique way.

God may be referred to as a Spirit, in that he is not material and that he transcends everything physical or even spatial. This does not mean that God is something vague or distant or impersonal. If God is our creator, then it is also reasonable to assume that he has a personality. Humans have personality and the Bible says God sometimes appears in the same image as the humans he created. God may be many things but first and foremost, he is a person. To say that God is anything less is to make him nothing more than the space between all matter.

WHAT ARE THE ATTRIBUTES OF GOD?

No single chapter in any one book can even come close to being comprehensive on the subject of the Godhead. The reader is encouraged to consider the many passages in scripture that reveal his varied manifestations and sides. Also, I believe it is wise to compare the Christian example of God with those of other religions. According to the Bible, these are but a few of the many examples.

OMNIPOTENCE, OMNISCIENCE & OMNIPRESENCE

The boundlessness of God is expressed in these three main attributes, which are inter-related. Consistent with the nature of his will, God is able to do everything. He is omnipotent (Gen. 17:1, Ex. 6:3, Job 5:17, Psa. 68:14, Matt. 19:26, Mk. 10:27).

Revelation 19:6
"And I heard as it were the voice of a great multitude,
and as the voice of many waters, and as the voice
of mighty thunderings, saying, Alleluia: for
the Lord God omnipotent reigneth."

Also God is omniscient. He knows everything instantaneously and eternally. His knowledge is completely comprehensive and totally devoid of ignorance. There is absolutly nothing that God does not know about and he never forgets anything. To say that God knows everything about everything is an understatement because humans are too mortal to fully appreciate such a supernatural concept (Job 12:22, Job 36:6, Psa. 33:13, Psa. 139:11).

Acts 15:18
"Known unto God are all his works from
the beginning of the world."

1 John 3:20
"For if our heart condemn us, God is
greater than our heart, and
knoweth all things."

The fact that God is omnipresent is because somehow God is everywhere, all the time permanently and somehow at once. There is nothing that God does not know because wherever it happens, he is there as a firsthand witness. Also in some way God is present in all points of space and time continually (I Kin. 8:27, 2 Chr. 2:6, Psa. 139:7, Isa. 66:1, Jer. 23:23, Acts 17:27). How this is possible is not within our human comprehension. We do not know because human knowledge is limited. Maybe we never can know while we are alive on this earth. King David knew the mystery of God's omnipresence when he wrote this Psalm.

Psalms 139:7
"Whither shall I go from thy spirit? or whither shall I flee from thy presence? v8 If I ascend up into heaven, thou art there: if I make my bed in hell, behold, thou art there. v9 If I take the wings of the morning, and dwell in the uttermost parts of the sea; v10 Even there shall thy hand lead me and thy right hand shall hold me."

GOD IS LOVE

Of all the attributes of God, one can be singled out as particularly descriptive. It alone, says so much about the nature and essence of God that in its own special way, it supersedes all others. God is Love. Just as we need blood to flow through our veins, love is the very breath of God. We are told that it is a vital part of his being. The love of God is manifested in the incarnation and atonement of his son Jesus Christ. He gave his son, to die on the cross, so that humanity might have a road to salvation. This single act, to Christians, reveals the full measure of his concern for humanity and all his creation. God loves us so much that he was willing to make a great personal sacrifice. One way that we can please God and know God is to love people and the things of God to the maximum of our ability.

1 John 4:8
"He that loveth not knoweth not God; for God is love."

GOD IS RIGHTEOUS

The righteousness of God is one of his primary attributes. This can easily start with his faithfulness to the covenants he has made. It continues with his zeal for the social rights of the poor and helpless. God has great compassion for all those who are in misery. The campassion of God has no bounds. He also shows his patience in dealing with those deserving of punishment and his impartiality of justice is clearly seen. But God also has a wrath which can be understood as an expression of his righteous aversion to sin. God hates all badness.

GOD IS ALL POWERFUL

When it comes to power, God's power is total and complete. The Bible refers to God as the "Lord God Almighty" (Gen. 17:1, Ex. 6:3). Even in the last book of the New Testament God makes a point to describe himself in no uncertain terms.

Revelation 1:8
"I am Alpha and Omega, the beginning and the ending, saith the Lord, which is, and which was, and which is to come, the Almighty."

GOD WANTS US TO SEEK HIS FACE

How can we be certain that God wants and expects us to seek his face? It is because that's exactly how God worded it when he appeared to Solomon one night long ago. Solomon had been praying for the forgiveness of Israel and God told him the requirements in:

II Chronicles 7:14
"If my people, which are called by my name, shall humble themselves, and pray, and seek my face, and turn from their wicked ways; then will I hear from heaven, and will forgive their sin, and will heal their land."

GOD CREATED EVERYTHING

God is shown as the creator and the beginning of all things in many passages of scripture. A Bible study on this subject alone can be particularly insightful (Gen. 1:1, Gen. 1:27, Psa. 19:1, Psa. 89:11, Isa. 37:16, Isa. 45:8, Neh. 9:6). In the book of John, the apostle proclaims that:

John 1:3
"All things were made by him, and without him was not anything made that was made."

John's statement is unmistakably conclusive. He wants the reader to know the full abilities and condition of the Godhead. Also, we see that his point leaves no holes as to the power and abilities of Almighty God. The God, John speaks of, has no shortcomings or limitations. The matter is brought home even further in:

Colossians 1:16
"For by him were all things created, that are in heaven, and that are in earth, visible and invisible, whether they be thrones, or dominions, or principalities, or powers: all things were created by him, and for him: v17 And he is before all things, and by him all things consist."

THE IMPARTIALITY OF GOD

Some people wonder if God prefers certain individuals over others based on his own whim or prejudice. This is never the way God is, and it cannot be the case, because God is not a respecter of persons (Rom. 2:11, Acts 10:34, Gal. 2:6). God is the holder of a perfectly holy balance that makes him the fairest person imaginable.

Ephesians, 6:8
"Knowing that whatsoever good thing any man doeth, the same shall he receive of the Lord, whether he be bond or free."

THE PERMANENCE OF GOD

God does not change. From the beginning, God chose to be this way. He is the same yesterday, today and forever. In our world, everything is undergoing constant change. But God never changes. The immutability of God is one of his most important identifiers. If God could or would change then that means he might change his mind. If God might change his mind about anything, then it could include his salvation plan. But his plan never changes (Num. 23:19, 1 Sam. 15:29, Job 23:13, Psa. 33:11).

Micah 3:6
"For I am the Lord, I change not; therefore
ye sons of Jacob are not consumed."

James 1:17
"Every good gift and every perfect gift is from above,
and cometh down from the Father of lights,
with whom is no variableness, neither,
shadow of turning."

GOD IS UNKNOWN

God is also good, merciful, faithful, true, holy, just, and probably the most important of all, unknown. This last point is important because with all the information in the Bible about God, he still is in fact unknown (Job 37:5, Psa. 40:5, Isa. 45:15). We cannot even guess his age because he is not limited by our knowlege of numbers and time.

Job 36:26
"Behold, God is great, and we know him not, neither
can the number of his years be searched out."

Ecclesiastics l3:11
"He hath made every thing beautiful in his time: also
he hath set the world in their heart, so that no
man can find out the work that God maketh
from the beginning to the end."

Chapter 18

WHO IS JESUS CHRIST?

The only Jesus Christ ever referenced here, is the Jesus Christ in the Holy Bible. This may seem obvious to some people but not to others. Later, we will compare several passages to help in our study, but the reader should keep in mind one thing. Jesus Christ is described in many different ways throughout church history.

Christians know him as the Son of the living God who died on the cross to save humanity from sin and even death. As a rule, Muslims, although proclaiming Jesus as a great prophet, do not believe in his resurrection from the dead. Also, they do not believe that he is the Son of God or equal to God in any way.

Followers of Hinduism and Buddhism mostly consider Jesus among the greatest men that ever lived. Neither give exclusive adoration to Jesus. Most Jews do not believe that Jesus Christ is the Messiah or Savior, whom they believe has still not yet come to save them. Knowing the different ways that sincere scholars, from various religions see Jesus, is very important to this study. I believe that this discovery can aid researchers in learning why topographic images on Earth might be helpful in resolving religious and doctrinal differences.

While doing this, let us also look closer at the identity of Jesus Christ. There will be no attempt to be comprehensive. This subject is a lifetime project for any Christian. I cannot escape my own Protestant bias, but I refuse to attack or even criticize any other religion or faith in this book. The majority of Christians, Catholics, Protestants and others believe in most of the basic tenets of the faith. This is encouraging.

WHAT DOES THE BIBLE SAY ABOUT JESUS?

I am including some facts on what the Bible says about Jesus because I do not want to be misunderstood. My own belief about the identity of the young king in prayer, outlined in the topography of the Middle East, is biased. I am resolved that the evidence on the Earth's own topography requires related information that is only provided in the Holy Bible. Then the evidence becomes really compelling. I think it would be wrong not to provide at least some of the basics.

Books have been written about Jesus as a mystic, a savior, an angel and Jesus as God. Debates abound concerning the trinity of God and the position of Jesus in the Godhead. So many ideas and opinions are enough to scare away any novice interested in the study of Jesus. This should not be the case. Most Christian religions divide the theological issues into those that are crucial and those that are not. For example, the virgin birth, the crucifixion and the resurrection of Jesus are basic beliefs for most Christians. These are considered holy truths and fundamental tenets of the faith. Such matters as infant baptism or sprinkling versus submersion are often left to more philosophical debate. The main agreement is that God's son, Jesus Christ, died for the sins of all *repentant* humanity.

Space limits a comprehensive study here and only careful Bible study is adequate to meet the subject. Still, these are some of the most important areas of what most Christians believe the Bible teaches about Jesus. I encourage serious Bible study on the subject of Jesus and the Godhead. In my mind, the best spiritual place to start is by joining a church near your home.

I will not single out any specific church preference, because what is most important, is how these churches teach and what you learn. Whether your choice is to join the Methodists, Baptists, Presbyterians, Catholics or any other Bible believing church, it should be your choice. If you become *hungry* for knowledge about God and *look first to the Bible* for your answers, you can not go wrong. Church doctrine, although very important, is fluid and subject to change. A church is important, but never to the point of demanding your loyalty outside the Biblical doctrine of God and Christ. What is most vital is *our personal relationship with God.*

IS JESUS CHRIST THE SON OF GOD?

Before citing various scriptures, it should be kept in mind that all the information about Jesus in the Bible, really boils down to one question. Is Jesus Christ the Son of God? Be careful though, by saying he is the Son, is to make him the same as God the Creator. That is why this question means everything. All of the other questions and facts are auxiliary and unless he is the Son of God, all of his promises are worthy of legitimate scorn.

THE HUMANITY OF JESUS CHRIST

After being told that Jesus is God incarnate, it is often difficult for some people to believe that he was also human, but he was. The prophet Isaiah wrote that the Messiah would be born of a virgin (Isa. 7:14). Jesus Christ had a human soul and a human spirit (Matt. 26:38; Lk. 23:46). He was born of a woman (Gal. 4:4), subject to growth (Lk. 2:52) and lived and worked among men (1 Jn. 1:1; Matt. 26:12). Also, Jesus was sinless (Heb. 4:15)and he was subject to the limitations of being human. Jesus was hungry (Matt. 4:2) and thirsty (Jn. 19:28) and knew what it was like to be tired (Jn. 4:6). To make things complete, just like God, Jesus wept (Jn. 11:35). Jesus came as the Son of Man (Lk. 19:10) and as the Messiah, also the son of David (Mk 10:47). Both the Old and New Testaments speak of Jesus as a man (Isa. 53:3; 1 Tim. 2:5). Jesus Christ lived as a man and he also died as one. It was the part of him that was also God that led to his resurrection. This was the ransom and the payment, so that all of humanity might have a guaranteed chance to be saved.

THE DEITY OF JESUS

After the scriptural evidence of the humanity of Jesus, it is important to know that he was and is also God. He is the creator of Heaven and Earth and his coming to the Earth, as a human, is part of the Salvation Plan. In explaining Christ as superior to the angels, quoting the Psalms, the book of Hebrews also proclaims the deity of Jesus Christ. The passage says that he is God and that his throne is for ever and ever.

Hebrews 1:8-10

"But unto the Son he saith, Thy throne, O God, is for ever and ever: a scepter of righteousness is the scepter of thy kingdom. v9 Thou hast loved righteousness, and hated iniquity; therefore God, even thy God, hath anointed thee with the oil of gladness above thy fellows. v10 And, Thou, Lord, in the beginning hast laid the foundation of the earth; and the heavens are the works of thine hands:"

THE ATTRIBUTES OF JESUS

Whenever we consider the attributes of God, they also apply to Jesus (Heb. 13:8). Jesus as the savior of humanity is no myth. He is the "truth (Jn. 14:6) and the life" (Jn. 1:4). Jesus has characteristics that identify his deity. He is omnipotent (Matt. 28:18) and omnipresent (Matt. 18:20). He even has the power to send the Holy Spirit as the comforter of all humanity (Jn. 15:26). The Bible shows us that angels worship Jesus (Heb. 1:16). So do men (Matt. 14:33) and all people and more (Phil. 2:10). We know that as part of the Godhead, he is equal to God the Father (Jn. 10:30; 14:23). Jesus is described as the Creator of everything (Jn. 1:3) who can forgive sins (Lk. 7:48), raise the dead (Jn. 5:25) and as the king who sits in judgment of all humanity.

John 5:27-28
"For as the Father hath life in himself; so hath he given to the Son to have life in himself; v27 And hath given him authority to executed judgment also, because he is the Son of Man."

WHY DID JESUS COME TO EARTH?

The main reason that Jesus Christ came to earth, as a man, was to provide a sacrifice for the sins of mankind (Heb. 10:1-10). He knew in advance what pain he would suffer. Also, it was his purpose to reveal God to humanity (Jn. 1:18) in his own unique way. Jesus knew that by coming to Planet Earth and saving us from our sins, he was also solving another major problem. He was destroying the works of the Devil (1 Jn. 3:8). Acting as a High Priest full of mercy (Heb. 5:1-2) Jesus was fulfilling the covenant of King David (Lk. 1:31-33). It was meant for Jesus to be the focus of exaltation (Phil. 2:9). Although so much of this reality about the Son of God is still a mystery, Jesus Christ is seen in the Bible as the perfect example of perfect humanity and undiminished deity, united in one person, forever.

In the book of John there is a story in the sixth chapter about Jesus telling the people exactly why he came to Planet Earth. The people there had seen him work a number of miracles and many of them believed that he was a true prophet of God. This was the first time that Jesus made it clear exactly who he really was, God incarnate, and why he came to Earth. It was in this speech that Jesus tells the people that it was not his choice that he came from Heaven as the Savior, but it was the *will* of his Father. It is his way of informing us of his duty. This is an obedient king. He is doing this not by his own choice but because of the will of his father. His sacrifice is important to everyone because Jesus also says that it will be he alone that will be in a position to provide eternal life in the last days. Here he speaks in:

John 6:38-40
"For I came down from heaven, not to do mine own will, but the will of him that sent me. v39 And this is the Father's will which hath sent me, that of all which he hath given me I should lose nothing, but should raise it up again at the last day. v40 And this is the will of him that sent me, that every one which seeth the Son, and believeth on him, may have everlasting life: and I will raise him up at the last day."

Some of the people there that day were religious leaders who mocked Jesus and murmured against him (v41) for making himself equal to God. To them, this was a serious, unforgivable blasphemy. These people knew him and his mother and father (v42). How could he dare claim his own deity? After Jesus made his proclamation, the people were so startled that many of his disciples stopped following him (v66) that very day.

Some people want to understand and compare Jesus as God and Jesus as man. In truth, the only logical conclusion anyone can reach through Bible study, is that Jesus had some limitations while he served his purpose in the form of a man. What were the limitations of the incarnate Christ on Earth? This interesting question is often referred to as "The Kenosis."

One false theory of Kenosis is that Christ gave up certain attributes during his earthly life. If this were so, then he ceased to be God during that time. Of course, an all powerful God could limit an incarnation of himself if he chose. Much of this debate, on whether Jesus was limited voluntarily or involuntarily, is really besides the point. We know that he condescended to taking on the likeness of sinful flesh in the incarnation. We also know that he showed more human-like than God-like attributes in his early life.

THE LAMB OF GOD

If the images in this map study are symbolic clues, which help us discover the identity of each person shown, then the elements are important. But once the story begins to make logical sense, we must then look closer at the possible meanings of the metaphor. Whether the facial features of the king in prayer look like the traditional statues and paintings of Jesus or of that of a lamb is significant. Also, the head and face being similar to a nappy wool-faced lamb (see pages 173-175) is appropriate. Jesus Christ is known as the "Lamb of God." The phrase "Lamb of God" is in the New Testament twice, both appearing in the gospel of John. The first time is when John proclaims Jesus Christ as the Messiah and the savior of all humanity.

John 1:29
"The next day, John seeth Jesus coming unto him, and saith, Behold the Lamb of God, which taketh away the sins of the world."

The next day John was standing with two of his disciples and advised them to follow Jesus by making the same pronouncement. Upon his declaration, both disciples immediately leave John and follow Jesus Christ.

John 1:35-36
"Again the next day after John stood, and two of his disciples. v36 And looking upon Jesus as he walked, he saith, Behold the Lamb of God."

The apostle Peter considered Jesus the perfect sacrificial lamb sent to purify the souls of all repentant humanity.

1 Peter 1:19-20
"But with the precious blood of Christ, as of a lamb, without blemish and without spot: v20 Who verily was forordained before the foundation of the world, but was manifest in these last times for you."

In the book of Acts, Phillip uses similar terms when reading to an Ethiopian eunuch from the book of Isaiah. Philip compares Jesus Christ to a sheep that is led to the slaughter.

Acts 8:32
"The place of the scripture which he read was this, He was led as a sheep to the slaughter and like a lamb dumb before his shearer, so opened he not his mouth."

There are many more related references to the Lamb of God or Jesus as the Lamb of God but we must not forget that God has many attributes.

THE NAIL SCARRED HAND

I have had some people ask me an interesting question. Doesn't the Bible say that Jesus had nail scared hands? (Jn. 20:25-27.) The seeming discrepancy of the young king in prayer having a nail going through his wrist is really not a discrepancy at all. The Greek meaning of the word "hand" (ceir) as used in the New Testament means almost any area from the base of the elbow to the hand.[1] Also, most experts in anatomy agree that it is physically unlikely that a man could hold up his own weight with only a nail going through the soft cartilage of the palm. The more practical location would be the wrist. Also, there is substantial archaeological evidence showing that the wrist and feet are where nails were used. Included is the evidence from the burial cloth of the much debated Shroud of Turin.[2]

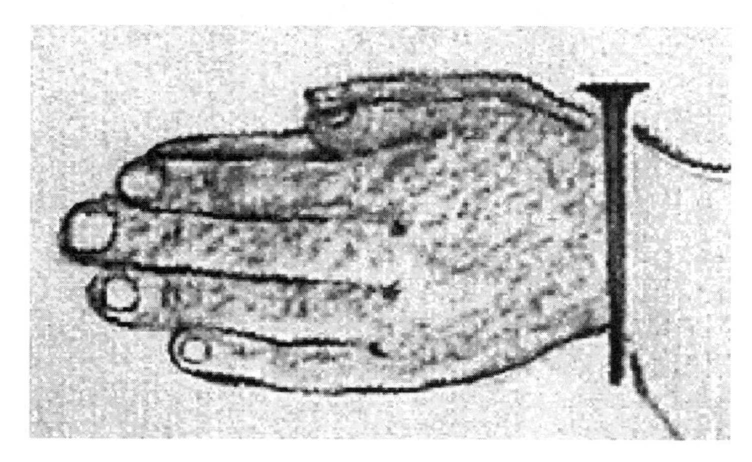

What Jesus did and any limitations he may have been subject to are irrelevant to the fact that Jesus is Savior and God. He has always had the power to establish any limitations upon himself that he might choose. Any speculations we have cannot negate or dispute his deity.

KING JESUS

The book of Revelation calls him the "King of Kings and Lord of Lords" (Rev. 19:16). According to the Bible, Jesus is and has always been the king of Heaven and the Lord of all creation. Jesus as king, is our sympathetic High Priest in Heaven (Heb. 4:14-16). He is designated as the head of the Church (Eph. 1:19-22) and impeccable in every way.

Although Jesus is the King of Kings, during his earthly ministry, he came as a servant to God for all humanity. He humbled himself as a servant and became our ransom for sin. Christ paid the price and penalty for our sins (Matt. 20:28; 1 Tim. 2:6). When Jesus Christ died on the cross, it was the culmination of the purpose of his incarnation (Matt. 20:28; Heb. 2:14). Jesus suffered the agony of the cross because of obedience to his Father. God the Father loved humanity so much that he made his Son a sacrifice. God had no scapegoat like Abraham, so he gave his Son. Just before Jesus is crucified, Pontis Pilate asked him if he was a king. Jesus responds by saying:

John 18:37
"...Thou sayest that I am a king. To this end was I born, and for this cause came I into the world, that I should bear witness unto the truth. Every one that is of the truth heareth my voice."

The world was changed by Christ's death so that all men might be saved (2 Cor. 5:18-19). In a way that we will not completely understand until later, our faith and belief, along with our love of God are necessary factors. According to the Holy Bible, Jesus served as the substitute for all of our sins, serving in our place (2 Cor. 5:21). Some people do not like to hear about Christ weeping and suffering but it is the truth. Jesus willingly permitted himself to suffer and die and be raised from the dead to show how much God loves us (Rom. 5:8).

JESUS RESURRECTED FROM THE DEAD

The reader is encouraged to study the events surrounding his resurrection in the Gospels of the New Testament (Matthew, Mark, Luke and John). This is because, if his death on the cross proves what kind of man he was, then his resurrection proves what kind of God he is. Also, such an impossible act as coming back from the dead, is certainly one thing that enables us to see the power of God.

The resurrection of Jesus was witnessed by many people. After the stone was rolled away from the tomb, Jesus appeared to Mary Magdalene (Jn. 20:11-17) and other women (Matt. 28:9-10). Later he appeared to Peter (1 Cor 15:5) and to some of his disciples on the road to Emmaus (Lk. 24:13-35). Also after appearing to the twelve disciples, he appeared to more than five hundred people at one time (1 Cor. 15:6).

THE IMPORTANCE OF THE RESURRECTION

Certain facts surrounding the resurrection of Jesus are important, to better understand the nature of these events. When Jesus died on the cross and then came back from the dead, it was no magic act. When he went to the cross, it was with a normal and actual human body, just like the body used by you and me (Jn. 20:20). His body was later proven through identification, as the same one laid in the tomb (Jn. 20:25-29).

The resurrection of Jesus Christ confirmed the truth of everything he said (Matt. 28;6). Jesus knew that we are already condemned. His resurrection promised the assurance of an afterlife to all those who follow him (1 Cor. 15:20-22). Jesus promised that he would not have favorites in this regard. He promised that anyone who comes to him, "I will in no way cast out." This means they cannot be subject to condemnation.

John 6:37
"All that the Father giveth me shall come to me; and him that cometh to me I will in no wise cast out."

When I look at any map of the Middle East and see the shape of the crowned king kneeling, it reminds me of what Christians call the "Blessed Hope." That is the hope and promise that Jesus Christ will return to Earth

No Other Way

and save the world. It reminds me of the fact that the only certain hope anyone has for total and complete salvation, is through Jesus Christ. The Bible says that there is *no other way.*

Acts 4:10-12
"Be it known unto you all, and to all the people of Israel, that by the name of Jesus Christ of Nazareth, whom ye crucified, whom God raised from the dead, even by him doth this man stand here before you whole. (v11) This is the stone which was set at nought of you builders, which is become the head of the corner. (v12) Neither is there salvation in any other: for there is none other name under heaven given among men, whereby we must be saved."

Bibliography

No attempt has been made to list all of the works consulted in the preparation of this book. The references listed here are those found particularly helpful and are intended to serve as a guide to further reading in the areas and categories specified.

Alighieri, Dante. *Purgatory.* New York: Hurst & Company Publishers, 1910.

Allegro, John. *The Dead Sea Scrolls - A Reappraisal.* Alphapedia, New York: Random House, 1977.

Anderson, J.K. *Tales of Great Dragons.* Santa Barbara, Ca.: Bellerophon Books, 1980.

Armstrong, Karen. *A History of God The 4000-Year Quest of Judaism, Christianity and Islam.* New York: Alfred A. Knoph Inc., 1994.

Arndt, W. *Does the Bible Contradict Itself? A Discussion of Alleged Contradictions in the Bible.* St. Louis: Concordia Publishing House, 1976.

Asimov, Isaac. *Words from the Myths.* New York: The New American Library, Inc.,1977.

Barclay, William. *Jesus as They Saw Him.* Grand Rapids: William B. Eerdmans Publishing Company, 1978.

Barrett, David B. *World Christian Encylopedia.* New York: Oxford University Press, 1982.

Bateman, Earl. *The Ocean Floor, Guide and Tourbook to the Map.* Berkeley: Celestial Arts Publishing, 1992.

Bauval, Robert, and Adrian Gilbert. *The Orion Mystery - Unlocking the Secrets of the Pyramids.* New York: Crown Publishers, 1994.

Carol Beckwith. "Niger's Wodaabe: People of the Taboo." *The National Geographic Magazine* (Oct 1983): 483-509.

Berger, John. *Ways of Seeing.* London: The British Broadcasting Corporation, 1979.

Berrigan, Daniel. *Jesus.* Garden City, N.Y.: Doubleday & Company, 1973.

Bokser, Ben Zion, trans. *The Talmud - Selected Writings.* New York: Paulist Press, 1989.

Booth, William. *How to Find God.* Lindale, Tex.: Last Days Ministries, 1982.

_______. *Who Cares?* Lindale, Tex.: Last Days Ministries, 1984.

Bower, T.G.R. "The Visual World of Infants." *Scientific American.* (December 1966): 80.

Boykin, John. *The Baha'i Faith.* Downers Grove, Il.: InterVarsity Press, 1982.

Brenton, Sir Lancelot C.L. *The Septuagint With Apocrypha Greek and English.* Peabody Ma.: Hendrickson Publishers, 1992.

Brooks, Keith L, and Irvine Robertson. *The Spirit of Truth and the Spirit of Error. Chicago: The Moody Bible Institute, 1975.*

Brownrigg, Ronald. *Who's Who in the Bible: The New Testament.* New York: Bonanza Books, 1971.

Bruce, F.F., ed. *The International Bible Commentary.* Grand Rapids: Marshall Morgan & Scott Publications Ltd., 1986.

Budge, E.A. Wallis. *The Egyptian Book of the Dead: The Papyrus Of Ani.* New York: Dover Publications Inc., 1967.

_______. *Osiris & the Egyptian Resurrection, Vol. I.* N.Y.: Dover Publictions Inc., 1973.

_______. *Osiris & the Egyptian Resurrection, Vol. II.* N.Y.: Dover Publications Inc., 1973.

Bureau of Naval Personnel. *Basic Optics and Optical Instruments.* New York: Dover Publications Inc., 1969.

Bullinger, E.W. *Number in Scripture: Its Supernatural Design and Spiritual Significance.* Grand Rapids: Kregal Publications, 1979.

Carpenter, Lloyd S. *Scientific Proof of the Deliberate Supernatural.* Pasadena, Ca.: Spiral Film Productions, 1988.

Cathie, Bruce L. *The Bridge To Infinity.* Boulder, Co.: America West Publishers, 1989.

Chandler, Russell. "Signs and Wonders." *Los Angeles Times,* 7 August 1983, 1.

Childress, David Hatcher. *Anti-Gravity and the World Grid.* Stelle, Il.: Adventures Unlimited Press, 1987.

_______. *Lost Cities & Ancient Mysteries of South America.* Stelle, Il.: Adventures Unlimited Press, 1986.

_______. *Lost Cities of Ancient Lemuria & the Pacific. Stelle,* Il.: Adventures Unlimited Press, 1988.

_______. *Lost Cities of North & Central America.* Stelle, Il.: Adventures Unlimited Press, 1992.

Clouse, Robert G., Richard V. Pierard, and Edwin M. Yamauchi. *Two Kingdoms: The Church and Culture Through the Ages.* Chicago: Moody Press, 1993.

Cobb, Vicki. *How to Really Fool Yourself.* New York: J.B. Lippincott, 1981.

Coleman, James A. *Relativity for the Layman: A Simplified Account of the History, Theory and Proofs of Relativity.* New York: The New American Library, 1958.

Comay, Joan. *Who's Who in the Bible: The Old Testament and The Apocrypha.* New York: Bonanza Books, 1971.

Cook, Roger. *The Tree of Life: Image for the Cosmos.* N. Y.: Thames and Hudson, 1988.

Coomaraswamy, Ananda. *Introduction of Indian Art.* Adyar, India: The Theosophical Publishing House, 1956.

Dawood, N.J., trans. *The Koran.* New York: Penguin Books, 1980.

De Laszlo, Violet Staub. *The Basic Writings of C.G. Jung.* New York: Random House, 1959.

Davidson, Gustav. *A Dictionary of Angels: Including the Fallen Angels.* New York: The Free Press, A Division of Macmillan, Inc., 1971.

Davidson, Marshall B., and Leonard Cottrell. *The Horizon Book of Lost Worlds.* New York: Doubleday & Company, Inc., 1962.

De Haan II, Martin R. *Bible Doctrine Summary: What's So Important About Doctrine?* Grand Rapids: Radio Bible Class, 1989.

_______. *Salvation: Position in Christ: What Does God Think of Me Now?* Grand Rapids: Radio Bible Class, 1987.

_______. *Salvation: Responsibility: What About Those Who Have Never Heard?* Grand Rapids: Radio Bible Class, 1988.

Dickey, Adam H. *God's Law of Adjustment.* Boston: The Christian Science Publishing Society, 1971.

Dickey, Thomas, Bance Muse, and Henry Wiencek. *Treasures of the World, the God Kings of Mexico.* Chicago: Stonehenge Press Inc., 1982.

Dubach, Harold W., and Robert W. Taber. *Questions About the Oceans.* Washington D.C., U.S. Naval Oceanographic Office, 1969.

Duncan, Ronald, and Miranda Weston Smith. *The Encyclopedia of Delusions.* New York: Simon & Schuster, 1979.

Earhart, Byron H. *Religious Traditions of the World.* San Francisco: Harper Collins Publishers, 1990.

Edgerton, Franklin, trans. *The Bhagavad Gita.* Cambridge, Ma.: Harvard University Press, 1972.

Eerdmans' Handbook to the Worlds Religions. Grand Rapids: Wm B. Eerdmans Publishing Company, 1982.

Eliot, Alexander. *Myths.* New York: McGraw Hill Book Company, 1976.

Elwell, Walter A. *Evangelical Commentary on the Bible.* Grand Rapids: Baker Book House, 1989.

Espenshade, Edward B. Jr., and Joel L. Morrison. *Goode's World Atlas, 16th Edition.* Chicago: Rand McNally & Company, 1978.

Estep, Howard C. *Why Do We Cry?* Colton, Ca.: World Prophetic Ministry, Inc., 1971.

Estes, Jane. "Lloyd Carpenter Sees The Face of God." *Pasadena Star News.* 27 Feb. 1983, B-1.

Evans, Craig A. *Noncanonical Writings and New Testament Interpretation.* Peabody Ma.: Hendrickson Publishers, 1992.

Fee, Gordon D., and Douglas Stuart. *How to Read the Bible for all its Worth.* Grand Rapids: Zondervan Publishing, 1993.

Ferguson, Everett. *Backgrounds of Early Christianity.* Grand Rapids: William B. Eerdmans Publishing Company, 1993.

Fournier, Lou and Merlinda. *Seven: An Historical and Cultural Overview.* Crofton, Md.: Second Nature Communications, 1988.

Fremantle, Francesca, and Chogyam Trungpa. (from the original by Guru Rinpoche according to Karma Lingpa). *The Tibetan Book of the Dead.* Boulder, Co.: Shambhala Publications, Inc., 1975.

Frick, Thomas. *The Sacred Theory of the Earth.* Berkeley: North Atlantic Books, 1986.

Froman, Robert. *Science, Art, and Visual Illusions.* New York: Simon and Schuster, 1971.

Frye, Northrop. *The Great Code: The Bible and Literature.* New York: Harcourt Brace Jovanovich, Publishers, 1989.

Gaebelein, A.C. *What The Bible Says About Angels.* Grand Rapids: Baker Book House, 1987.

Gaer, Joseph. *What the Great Religions Believe.* New York: Dodd, Mead & Co., 1963.

Gill, Sam D. *Native American Religions: An Introduction.* Belmont, Ca.: Wadsworth Publishing Company, 1982.

Glasser, William M.D. *Stations of the Mind: New Directions for Reality Therapy.* New York: Harper & Row Publishers, 1981.
Goetz, Delia, and Sylvanus G. Morley. *Popol Vuh, The Sacred Book of the Ancient Quiche Maya,* Norman Oklahoma, University of Oklahoma Press, 1969.
Goode's World Atlas. 16th Edition. San Francisco: Rand McNally & Company, 1982.
Graham, Billy. *Angels: God's Secret Agents.* New York: Pocket Books, 1975.
Grant, Michael. *Jesus, An Historian's Review of the Gospels.* New York: Charles Scribner's Sons, 1977.
Gray, Henry F.R.S. *Anatomy Descriptive and Surgical.* (Gray's Anatomy) New York: Bounty Books, 1977.
Grafton, Carol Belanger. *Optical Designs in Motion with Moire Overlays.* New York: Dover Publications Inc., 1976.
Green, Jay P. Sr., *The Interlinear Bible, Hebrew-Greek-English.* Peabody Mass: Hendrickson Publishers, 1986.
Green, Joel B., Scot McKnight, and Howard Marshall. *Dictionary of Jesus and The Gospels.* Downers Grove, Il.: InterVarsity Press, 1992.
Gregory, Richard L. "Visual Illusions." *Scientific American.* (November 1968): 66.
Halevi, Z'ev ben Shimon. *Kabbalah: Tradition of Hidden Knowledge.* New York: Thames and Hudson, 1979.
Hall, Manly P. *The Blessed Angels.* Los Angeles, Ca.: The Philosophical Research Society Inc., 1980.
_______. *Man, The Grand Symbol of the Mysteries.* Los Angeles, Ca.: The Philosophical Research Society Inc., 1972.
_______. *The Secret Teachings of All Ages.* Los Angeles, Ca.: The Philosophical Research Society Inc., 1977.
_______. *Studies in Character Analysis.* Los Angeles, Ca.: The Philosophical Research Society Inc., 1980.
_______. *Wit and Wisdom of the Immortals.* Los Angeles, Ca.: The Philosophical Research Society Inc., 1987.
Hardy, Alister, Robert Harvie and Arthur Koestler, *The Challenge of Chance.* New York: Random House, 1978.
Harm, Frederick R. *How to Respond to...The Science Religions.* St. Louis, Missouri: Concordia Publishing House, 1981.
Hawthorne, Gerald F., and Ralph P. Martin. *Dictionary of Paul And His Letters.* Downers Grove, Il.: InterVarsity Press, 1993.
Heidel, Alexander. *The Gilgamesh Epic and Old Testament Parallels.* Chicago: The Universityof Chicago Press, 1973.
Hesse, Hermann. *Siddhartha.* Hilda Rosner, trans. New York: New Directions Publishing Corporation, 1957.
Hoagland, Richard C. *The Monuments of Mars: A City on the Edge of Forever.* Berkeley: North American Books, 1987.
Ittelson, W.H., and F.P. Kilpatrick. "Experiments in Perception." *Scientific American* (August 1951): 50-55.

Irving, Clive. *Sayings of the Ayatollah Khomeini.* New York: Bantam Books Inc., 1980.

Jung, C.G., *Analytical Psychology: Its Theory and Practice.* New York: Pantheon Books, 1968.

_______. *Synchronicity; An Acausal Connecting Principle.* New York: Pantheon Books, 1968.

_______. *The Undiscovered Self.* Boston: Little, Brown and Company, 1958.

Kaplan, Aryeh, Rabbi. *The Living Torah: A New Translation Based On Traditional Jewish Sources.* New York: Maznaim Publishing Corporation, 1981.

Kersten, Holger, and Elmar R. Gruber. *The Jesus Conspiracy: The Turin Shroud & The Truth About the Resurrection.* New York: Barnes & Noble Books, 1995.

Key, Wilson Bryan. *The Clam-Plate Orgy and Other Subliminal Techniques for Manipulating Your Behavior.* New York: The New American Library Inc., 1980.

Klossowski de Rola, Stanislas. *Alchemy: The Secret Art.* New York: Thames and Hudson, 1973.

Koestler, Arthur. *The Roots of Coincidence.* New York: Random House Publishers, 1978.

Lamy, Lucie. *New Light on Ancient Knowledge, Egyptian Mysteries.* New York: Thames and Hudson, 1981.

Larcher, Jean. *Geometrical Designs & Optical Art.* New York: Dover Publications, 1974.

Laurence, Richard LL.D., (translator) *The Book of Enoch.* Thousand Oaks, Ca: Artisan Sales, 1980.

Lehmann, Arthur C. and James E. Myers. *Magic, Witchcraft, and Religion: An Anthropological Study of the Supernatural.* Palo Alto and London: Mayfield Publishing Company, 1985.

Lewis James F. *Religious Traditions of the World.* Grand Rapids: Zondervan Publishing House, 1991.

Lochhaas, Philip H. *How to respond to...Islam.* St. Louis, Missouri: Concordia Publishing House, 1979.

_______. *How to respond to...The Eastern Religions.* St. Louis, Missouri: Concordia Publishing House, 1981.

_______. *How to respond to...The New Christian Religions.* St. Louis, Missouri: Concordia Publishing House, 1979.

Long, Max Freedom. *What Jesus Taught In Secret.* Marina del Rey, Ca.: DeVorss & Company, 1980.

The Lost Books of The Bible and The Forgotten Books of Eden. U.S.A. Newfoundland: World Bible Publishers, Inc., 1926.

Louis, David. *2201 Fascinating Facts.* New York: Greenwich House, 1983.

Lovett, C.S. *Beginning to Pray.* Baldwin Park, Ca.: Personal Christianity, 1975.

Lucas, Jerry, and Del Washburn. *Theomatics: God's Best Kept Secret Revealed.* New York: Stein and Day Publishers, 1980.

Luce, Willard. "Utah Elements Etch Great Stone Faces." *Country Extra (*May 1990): 23.

Machado, Luis Alberto. *The Right to be Intelligent.* U.S.A.: Pergamon Press, 1980.

Maclagan, David. *Creation Myths: Man's Introduction to the World.* New York: Thames and Hudson, 1977.

Martin, William C. *The Layman's Bible Encyclopedia.* Nashville, Tn.: The Southwestern Company, 1964.

Mauro, Phillip. *The Number of Man : The Climax of Civilization.* London: Morgan & Scott Ltd., 1910.

May, Herbert G., and Bruce M. Metzger. *The New Oxford Annotated Bible, Revised Standard Version.* New York: Oxford University Press, 1971.

Mears, Henrietta C. *What the Bible is All About.* Ventura, Ca.: Regal Books, 1980.

Medina, David. *God's Secret Weapon: The Ark of the Covenant.* New York: Global Communications, 1982.

Meyer, Marvin W., trans. *The Secret Teachings of Jesus: Four Gnostic Gospels.* New York: Random House, 1984.

Minkel, Walter Kafton. *Subterranean Worlds: 100,000 Years of Dragons, Dwarfs, the Dead, Lost Races & UFO's From Inside the Earth.* Port Townsend, Wa.: Loompanics Unlimited, 1987.

Mitchell, James, ed. *The Random House Encyclopedia.* 3rd Edition. New York: Random House Publishing, 1990.

Morey, Earl Wesley. *Search the Scriptures.* Vienna, Virginia: Agape Ministry, Inc., 1993.

Moore Hyatt. *The Alphabet Makers.* Waxhaw, North Carolina: The Museum of the Alphabet, 1991.

Morris, Richard. *The End of the World.* Garden City, N.Y.: Anchor Press Doubleday, 1980.

Munger, Robert Boyd. *My Heart Christ's Home.* Downers Grove, Il.: InterVarsity Press, 1954.

Neisser, Ulric. "Visual Search." *Scientific American (*June 1964): 94.

Nierenberg, Gerald I., and Henry H. Calero. *How To Read A Person Like A Book.* New York: Pocket Books, 1971.

The NIV Study Bible, New International Version. Grand Rapids: Zondervan Bible Publishers, 1984.

Norvell. *Amazing Secrets of the Mystic East.* West Nyak, N.Y.: Parker Publishing Co. Inc., 1980.

O'Donnell, Edith "The Face on Mars and the Extraterrestrial Connection." Whole Life Expo., Pasadena, Ca.: Wrkshp #456 February 18, 1991, 8pm.

Oster, Gerald and Yasunori Nishijima. "Moire' Patterns." *Scientific American* (May 1963): 54.

O'Reilly, Philip. *1000 Questions and Answers on Catholicism.* New York: Henry Holt and Company, 1956.

Paraquin, Charles H. *The World's Best Optical Illusions.* New York: Sterling Publishing Company Inc., 1987.

Pelton, Robert W., and Karen W. Carden. *In My Name Shall They Cast Out Devils.* Cranbury, New Jersey: A.S. Barnes and Company, 1976.

Phelan, M. *Handbook of All Denominations.* Nashville, Tn.: Cokebury Press, 1924.
Pomeroy, Elizabeth. *The Huntington: Library, Art Gallery, Botanical Gardens.* (Museum Catalogue) Covent Garden, London: Scala Philip Wilson, 1983.
Pratney, Winkle. *The Holy Bible: Wholly True.* Lindale, Tex.: Last Days Ministries, 1985.
Purce, Jill. *The Mystic Spiral: Journey of the Soul.* New York:Thames and Hudson, 1974.
Rasmussen, Carl G. *Zondervan NIV Atlas of the Bible.* Grand Rapids: Zondervan Publishing House, 1989.
Ravenhill, Leonard. *The Judgement Seat of Christ.* Lindale, Tex.: Last Days Ministries, 1983.
Reader's Digest Great World Atlas, First Edition. Pleasantville, N.Y.: The Reader's Digest Association Inc., 1969.
Riedel, Eunice, Thomas Tracy, and Barbara D. Moskowitz. *The Book of the Bible.* New York: Bantam Books, 1981.
Rock, Irvin, and Charles S. Harris. "Vision and Touch." *Scientific American.* (May 1967): 104.
Robertson, J.M. *Pagan Christs.* New York: Barnes & Noble Books, 1966.
Robinson, James, ed. *The Nag Hammadi Library.* New York: Harper & Row, 1977.
Rongstad, James L. *How to Respond To...The Lodge.* St. Louis, Missouri: Concordia Publishing House. 1979.
Ross, Hugh. *What is Christianity?* Sierra Madre, Ca.: Dr. Hugh Ross, 1980.
_______. *Genesis One: A Scientific Perspective.* Sierra Madre, Ca.: Wisemen Productions, 1983.
Ryrie, Charles Caldwell. *The Ryrie Study Bible.* Chicago, Il.: Moody Press, 1978.
Saunders, Dale E. *Mudra: A Study of Symbolic Gestures in Japanese Buddhist Sculpture.* New York: Pantheon Books Inc., 1960.
Sanders, E.P. *The Historical Figure of Jesus.* Middlesex, England: Penguin Books Ltd., 1993.
Sanders, N.K. *The Epic of Gilgamesh.* An English Version with an Introduction. New York: Penguin Books, 1982.
Schiffman, H.R. *Sensation and Perception, An Integrated Approach.* Rutgers, The State University. New York: John Wiley & Sons Inc., 1978.
Schillebeeckx, Edward. *Jesus, an Experment in Christology.* Hubert Hoskins, trans. New York: Vintage Books, A division of Random House, 1981.
Schultz, Ted. *The Fringes of Reason: A Whole Earth Catalog.* New York: Harmony Books, 1989.
Sell, Henry T. *Studies of the Four Gospels.* New York: The Fleming H. Revell Co., 1996.
Simon, Seymour. *The Optical Illusion Book.* New York: William Morrow and Company, 1984.
Sitchin, Zecharia. *The Stairway to Heaven.* New York: St. Martin's Press, 1980.

Sladek, John. *The New Apocrypha: A Guide to Strange Sciences and Occult Beliefs.* London, Toronto, Sydney. New York: Granada Publishing, 1978.
Smart, Ninian and Richard D. Hecht, *Sacred Texts of The World, A Universal Anthology.* New York: The Crossroad Publishing Company, 1984.
Smith, Bradley, and Wang Weng. *China: A History in Art.* New York: Doubleday & Company, 1978.
Smith, Huston. *The Religions of Man.* New York: Harper & Row, Publishers, 1965.
Smith, Morton. *Jesus the Magician.* San Francisco, Ca.: Harper & Row Publishers, 1978.
_______. *The Secret Gospel.* Clearlake, Ca.: The Dawn Horse Press, 1982.
Smyth, Piazzi. *The Great Pyramid: Its Secrets and Mysteries Revealed.* New York: Bell Publishing Co., 1978.
Steinsaltz, Adin. *The Essential Talmud.* New York: Basic Books Inc. Publishers, 1976.
Strong, James. *The Exhaustive Concordance of the Bible.* New York: Abingdon Press, 1974.
Stuart, George E., and Gene S. Stuart. *The Mysterious Maya.* Washington D.C.: National Geographic Society, 1983.
Swanson, Reuben J. *The Horizontal Line Synopsis of the Gospels, Greek Edition, Volume I, The Gospel of Matthew.* Hillsboro, N. C.: Western North Carolina Press Inc., 1982.
Talbott, Stephen L. *Velikovsky Reconsidered.* New York: Warner Books, 1976.
Tansley, David V. *Subtle Body: Essence and Shadow.* New York: Thames and Hudson, 1984.
The Teaching of Buddha. Tokyo Japan: Kosaido Printing Co., 1978.
The Times Atlas of World History. Maplewood, New Jersey: Hammond Inc., 1993.
Thomas, E. Llewellyn. "Movements of the Eye." *Scientific American* (August 1968): 88.
Thurlow, Gilbert. *Biblical Myths & Mysteries.* New York: Crescent Books, 1974.
Tompkins, Peter. *Secrets of the Great Pyramid.* New York: Penguin, 1978.
Vaughan, Ian. *Incredible Coincidence: The Baffling World of Synchronicity.* New York: Ballantine Books, 1979.
Vermes, G. *The Dead Sea Scrolls in English.* New York: Penguin Books, 1975.
Vine, W.E. *Expository Dictionary of New Testament Words.* Nashville, Tn.: Thomas Nelson, Publishers, 1980.
Von Franz, Marie-Louise. *Time: Rhythm and Repose.* New York: Thames and Hudson, 1978.
Voss, Gilbert L. *Oceanography.* Racine, Wis.: Western Publishing Company, Inc., 1972.
Watts, Alan W. *Myth And Ritual In Christianity.* Boston: Beacon Press, 1968.
Whiston, William. *Josephus: Complete Works.* Grand Rapid: Kregal Publications, 1981.
Wigram, George V. *The New Englishman's Hebrew Concordance.* Peabody, Mass.: Hendrickson Publishers, 1984.
Wilken, Robert L. *The Christians as the Romans Saw Them.* New Haven: Yale University Press, 1984.

Wilson, Peter Lamborn. *Angels: Messengers of the Gods.* New York: Thames and Hudson, 1980.

_______. *Angels.* New York: Pantheon Books, 1980.

Winter, Ralph D., and Roberta H. Winter. *The Word Study New Testament.* Wheaton, Il.: Tyndale House Publishers Inc., 1978.

Woolley, Leonard. *UR: The First Phases.* New York: Penguin Books, 1946.

Wrixon, Fred B. *Codes, Ciphers, and Secret Languages.* New York: Bonanza Books, 1981.

Yamauchi, Edwin. *Jesus, Zoroaster, Buddah, Socrates, Muhammad.* Don Mills, Ontario, Canada: InterVarsity Press, 1972.

Zangger, Eberhard. *The Flood From Heaven: Deciphering the Atlantis Legend.* New York: William Morrow and Company Inc., 1992.

NOTES

Introduction

[1]Jane Estes, "Lloyd Carpenter Sees The Face of God," *Pasadena Star News,* 27 February 1983, B-1.

[2]Lloyd S. Carpenter, *Scientific Proof of the Deliberate Supernatural,* 54 min. (Pasadena, CA: Spiral Film Productions, 1988), videorecording.

[3]Russell Chandler, "Signs and Wonders," *Los Angeles Times,* 7 August 1983, 1.

[4]*NBC Channel 4 News* (Los Angeles, CA: National Broadcasting Company, 4 January 1985), 5:55 PM.

[5]*Hammond Atlas of the World* (Maplewood, NJ: Hammond Inc., 1993), back dustcover.

Chapter 1.

[1]Gilbert L. Voss, *Oceanography* (Racine, WI: Western Publishing Co., Inc., 1972), 7.

[2]David Louis, *2201 Fascinating Facts* (New York: Greenwich House, 1983), 351.

[3]Edward B. Espenshade, Jr. and Joel L. Morrison, *Goode's World Atlas,* 16th ed., (San Francisco: Rand McNally Company, 1982), 233.

[4]Ibid.

[5]James Mitchell, ed., *The Random House Encyclopedia,* 3d ed. (New York: Random House Publishing, 1990), 230-235.

[6]Espenshade, 229.

[7]Ibid.

[8]Ibid.

[9]Voss., 6.

[10]Ibid.

[11]Ibid., 7.

[12]Ibid.

Chapter 2.

[1]*Alexander Eliot, et al., eds., Myths* (New York: McGraw Hill Book Co., 1976), 93.

[2]Lucie Lamy, *New Light on Ancient Knowledge: Egyptian Mysteries* (New York: Thames and Hudson 1981), 15.

[3]Delia Goetz and Sylvanus G. Morley, *Popol Vuh, The Sacred Book of the Ancient Quiche Maya* (Norman, OK: University of Oklahoma Press, 1969), 138.

[4]Manly Hall, *Man, The Grand Symbol of the Mysteries* (Los Angeles:The Philosophical Research Society Inc., 1972), 217.

[5]Ibid., 228.

[6]Psalm 17:8. All scripture quotations are from the King James Version unless otherwise indicated.

[7]Ezekiel 7:4.

[8]I Corinthians 15:52.

[9]Dante Alighieri, *Purgatory* (New York: Hurst & Co. Publishers, N.D.), Canto II., p. 241.

[10]Ibid., Canto VI., p. 264.

Chapter 3.

[1]G. Vermes, *The Dead Sea Scrolls in English* (New York: Penguin Books 1995), Hymn IX., 179.

[2]Manly P. Hall, *The Secret Teachings of All Ages* (Los Angeles: The Philosophical Research Society Inc., 1977), LXXIX.

[3]N.K. Sanders, *The Epic of Gilgamesh* (New York: Penquin Books, 1982), 110.

[4]James Mitchell, ed. "Eye," *Random House Encyclopedia*, 3d ed. (New York: Random House, 1990), 2178.

[5]Lamy, 12.

[6]E.A. Wallis Budge, *The Egyptian Book of the Dead, The Papyrus Of Ani* (New York: Dover Publications, Inc., 1967), back cover.

[7]E.A. Wallis Budge, *Osiris & the Egyptian Resurrection.,* vol. I (New York: Dover Publications, Inc., 1973), 102.

[8]Ibid., 104.

[9]Ibid., Vol. II, p. 38.

[10]Louis., 26.

[11]Ibid., 350.

[12]Gen. 23:2.

[13]Gen. 27:38.

[14]Gen. 29:11.

[15]Gen. 33:3.

[16]Gen. 50:17.

[17]Jer.13:16.

[18]Lk. 7:38.

[19]David Hatcher Childress, *Lost Cities and Ancient Mysteries of South America* (Stelle, Il.: Adventures Unlimited Press, 1986), 145.

[20]James Strong, *The Exhaustive Concordance of the Bible* (Nashville-New York: Abingdon Press, 1974), 14.

Chapter 4.

[1]Elizabeth Pomeroy, *The Huntington Library Art Gallery., Museum Catalogue* (Covent Garden, London: Philip Wilson Publishers Ltd., 1983), 66.

[2]Ibid., 62.

[3]Ibid., 74.

[4]"Omni Competition" *Omni* (April 1985): 106.

[5]C.G. Jung, *Synchronicity; An Acausal Connecting Principle,* The Interpretation of Nature and the Psyche, (New York: Pantheon Books, 1955).

[6]Violet Staub De Laszlo, *The Basic Writings of C.G. Jung* (New York: Random House, 1959), 101.

[7]Arthur Koestler, *The Roots of Coincidence* (New York: Random House, 1972).

[8]Alister Hardy, Robert Harvie and Arthur Koestler, *The Challenge of Chance* (New York: Random House, 1974).

[9]Alan Vaughan, *Incredible Coincidence, The Baffling World of Synchronicity* (New York: Ballantine Books 1979), 3.

[10]W.H. Ittelson and F.P. Kilpatrick, "Experiments in Perception," *Scientific American* (August 1951): 50.
[11]Richard L. Gregory, "Visual Illusions," *Scientific American* (Nov. 1968): 66.
[12]Ulric Neisser, "Visual Search," *Scientific American* (June 1964): 94.
[13]Gerald Oster and Yasunori Nishijima, "Moire' Patterns," *Scientific American* (May 1963): 54.
[14]E. Llewellyn Thomas, "Movements of the Eye." *Scientific American* (August 1968): 88.
[15]T.G.R. Bower, "The Visual World of Infants" *Scientific American* (Dec. 1966): 80.
[16]Irvin Rock and Charles S. Harris, "Vision and Touch," *Scientific American* (May 1967): 104.
[17]H.R. Schiffman, *Sensation and Perception, An Integrated Approach* (New York: Rutgers, The State University, 1978), 287.
[18]Thomas Dickey, Bance Muse, and Henry Wiencek, *Treasures of the World, The God Kings of Mexico* (Chicago: Stonehenge Press Inc., 1982), 43.
[19]Carol Beckwith, "Niger's Wodaabe: People of the Taboo" *The National Geographic Magazine* (Oct 1983): 483-509.
[20]Bureau of Naval Personnel, *Basic Optics and Optical Instruments* (New York: Dover Publictions, Inc., 1968), 115.
[21]Lamy, 16.
[22]Bradley Smith & Wan-go Weng, *China, A History in Art,* (New York: Doubleday & Company, 1978), 75.
[23]Carpenter.

Chapter 5

[1]James Mitchell, ed. "Skin and Hair," *Random House Encyclopedia,* 3d ed. (New York: Random House, 1990), 687.

Chapter 6

[1]Joseph T. Shipley, *Dictionary of Word Origins* (New York: Dorset Press, 1955), 148.
[2]Hall, *The Secret Teachings of All Ages,* CXXXIII.
[3]Ibid., XIX.
[4]Strong, "Hebrew and Chaldee Dictionary," *Exhaustive Concordance* (Nashville:Abingdon Press, 1974), 77.
[5]Hall, *Man, Grand Symbol of the Mysteries:Thoughts in Occult Anatomy* (Los Angeles: Philosophical Research Society, 1972), 226.
[6]E. A. Wallis Budge, *An Egyptian Hieroglyphic Dictionary* (New York: Dover Publications Inc.), 936.
[7]Ibid., LXXIII.
[8]Ex. 20:4.

Chapter 7

[1]Strong, *Concordance,* 910.

[2]Ibid., "Dictionary," 112.
[3]Winkle Pratney, *The Holy Bible Wholly True* (Lindale, TX: Last Days Ministries, 1983), 4 & 5.
[4]Ibid.
[5]E.W. Bullinger, *Number in Scripture: Its Supernatural Design and Spiritual Significance* (Grand Rapids: Kregal Publications, 1979), 168.
[6]Ibid., 170.
[7]Ibid., 173.
[8]Ibid., 177.
[9]Lou and Merlinda Fournier, *Seven: An Historical and Cultural Overview* (Crofton, MD: Second Nature Communications, 1988), 37.
[10]Ibid.
[11]Ibid.
[12]Strong, 910.
[13]Fournier, 37.
[14]Ibid.
[15]Ibid.
[16]Hall, *The Secret Teachings of All Ages*, CLXXXV.
[17]Ibid., LXV.
[18]Henry Gray F.R.S., *Anatomy Descriptive and Surgical,* (New York: Bounty Books, 1977), 81.
[19]Fournier, 34.
[20]Gray, 643.
[21]Ibid.
[22]Ibid., 37.
[23]Ibid., 17.
[23]Bruce L. Cathie, *The Bridge To Infinity,* (Boulder, CO: America West Publishers, 1989), 35.
[24]Ibid., 36.
[25]Ibid.
[26]Ibid.
[27]Ibid.
[28]Ibid.
[29]Ibid.
[30]Lamy, 12.
Chapter 8
[1]Timothy Green Beckley, *The Visitation, Modern Miracles & Signs* (New York: Global Communications, 1981), 1.
[2]Wilson Bryan Key, New York: *The Clam-Plate Orgy and Other Subliminal Techniques for Manipulating your Behavior* (New York: The New American Library Inc., 1980), 114-115.
[3]"Mt. St. Helen's 8X10 Photo" (Gravette, AK: The Shepherd's Chapel, 1992),

Catalogue Item # 38.00.

[4]*Los Angeles Herald Examiner,* "Bakersfield 'Miracle' Still Drawing Faithful" (January 7, 1985), A3 col.5.

[5]*Weekly World News,* "Face Of Jesus Photographed Over Somalia" (March 2, 1993), 24.

[6]Willard Luce, "Utah Elements Etch Great Stone Faces" *Country Extra* (May 1990), 23.

[7]Matt. 27:33, Mk. 15:22, Jn. 19:17.

[8]"Skeptical Eye Facing Up to Mars,"*Discover* (Chicago, IL: April 1985): 92.

[9]Vincent Di Pietro, and Gregory Molenaar, *Unusual Martian Surface Features,* Third Edition (Glendale Md: Mars Research 1982), 23.

[10]Richard C. Hoagland, *The Monuments of Mars: A City on the Edge of Forever* (Berkeley, CA: North American Books, 1987), Illus. #10, p. 222.

[11]Ibid., Illus. #27, p. 239.

[12]Ibid., Illus. #3033, p. 242 & 243.

[13]*Discover*, 92.

[14]Ibid.

[15]Ibid.

[16]Edith O'Donnell, "The Face on Mars and the Extraterrestrial Connection," Whole Life Expo., Pasadena, CA, Wrkshp #456 February 18, 1991, 8pm.

[17]Carpenter.

[18]K.C. Cole, "Rock May Bear Signs of Ancient Life on Mars," *Los Angeles Times,* 7 August 1996, 1& 8.

Chapter 9

[1]G. Vermes, *The Dead Sea Scrolls in English,* (New York, Penguin Books 1995), Hymn IX., p. 160.

[2]Walter Kafton-Minkel, *Subterranean Worlds: 100,000 Years of Dragons,Dwarfs, the Dead Lost Races & UFO's From Inside the Earth* (Port Townsend, WA: Loompanics Unlimited, 1989), 8.

Chapter 10

[1]Strong, 281.

[2]W.E. Vine, *An Expository Dictionary of New Testament Words* (Nashville: Thomas Nelson Inc., 1980), 328.

[3]Lechtman.

[4]J.K. Anderson, *Tales of Great Dragon* (Santa Barbara, CA: Bellerophon Books, 1980), 3.

[5]Charles Caldwell Ryrie, *The Ryrie Study Bible* (Chicago: Moody Press, 1978), 712.

[6]Ibid., 1063, 1077.

[7]Ibid., 1270.

[8]Ibid., 1049.

[9]Kafton-Minkel, 42 & 43.

[10]Hall, *The Secret Teachings of All Ages,* LXXXVI.

[11]Walter A. Elwell, *Evangelical Commentary on the Bible* (Grand Rapids: Baker Book House, 1989), 1216.

Chapter 11

[1]Elwell, 6.

Chapter 18

[1]Strong, "Dictionary," 77.

[2]Holger Kersten and Elmar R. Gruber, *The Jesus Conspiracy* (New York: Barnes & Noble Books, 1992), 288.

ILLUSTRATION ACKNOWLEDGMENTS

All photographs not otherwise credited were taken by the author.

Page

6. National Oceanic and Atmospheric Administration.
8. Rand McNally Map Company
9. Map, Rand McNally Map Company, inset, Anna Carpenter
10. Rand McNally Map Company
11. Rand McNally Map Company
12. Rand McNally Map Company, tracing, Anna Carpenter
13. Rand McNally Map Company
17. Both, Rand McNally Map Company
18. Both, Rand McNally Map Company
19. Both, Rand McNally Map Company
20. Rand McNally Map Company
21. Rand McNally Map Company
22. Anna Carpenter
27. Map, Rand McNally Map Company, inset, Anna Carpenter
28. Map, Rand McNally Map Company, inset, Anna Carpenter
29. National Oceanic and Atmospheric Administration.
33. From *Lost Cities and Ancient Mysteries of South America*, Adventures Unlimited Press.
37. Anna Carpenter
39. Top by author, bottom from ancient woodcut.
40. Anna Carpenter
41. Enlargement of small drawing by Louis (last name unknown), Cannes, France.
45. Author
47. From *Basic Optic and Optical Instruments,* Bureau of Naval Personnel.
48. Photo by author.
53. From *Man, Grand Symbol of the Mysteries Thoughts in Occult Anatomy,* The Philosophical Research Society Inc.
55. From *Man, Grand Symbol of the Mysteries Thoughts in Occult Anatomy,* The Philosophical Research Society, Inc.
57. Top From *Man, Grand Symbol of the Mysteries Thoughts in Occult Anatomy,* The Philosophical Research Society, Inc., bottom, Rand McNally Map Company.

58. Top From *Man, Grand Symbol of the Mysteries Thoughts in Occult Anatomy,* The Philosophical Research Society, Inc. bottom, Rand McNally Map Company
59. Left From *Man, Grand Symbol of the Mysteries Thoughts in Occult Anatomy,* The Philosophical Research Society, Inc., bottom, Right, Rand McNally Map Company
60. Rand McNally Map Company
61. Left, From *Man, Grand Symbol of the Mysteries Thoughts in Occult Anatomy,* The Philosophical Research Society, Inc., bottom, Right, Rand McNally Map Company
63. Reproduction of letter by Rabbi Alan R. Lachtman.
65. From, *The Secret Teachings of All Ages,* The Philosophical Research Society, Inc.
66. From, *The Secret Teachings of All Ages,* The Philosophical Research Society, Inc.
69. From, *Man, Grand Symbol of the Mysteries Thoughts in Occult Anatomy,* The Philosophical Research Society, Inc.
70. Photo by author.
74. Rand McNally Map Company
75. Rand McNally Map Company
82. From, *The Secret Teachings of All Ages,* The Philosophical Research Society, Inc.
87. From, *The Secret Teachings of All Ages,* The Philosophical Research Society, Inc.
88. Celestial Arts, Berkeley, Ca.
89. Celestial Arts, Berkeley, Ca.
98. NASA
105. Top, Rand McNally Map Company, bottom, Anna Carpenter
106. Anna Carpenter
107. Rand McNally Map Company
109. Both, Rand McNally Map Comapany
110. Anna Carpenter
114. Anna Carpenter
115. Rand McNally Map Company
117. From, *Subterranean Worlds 100,000 Years of Dragons, Dwarfs, the Dead Lost Races & UFO's from Inside the Earth,* Loompanics Unlimited, Port Townsend, Wa.
118. Both, Rand McNally Map Company
119. Both, Rand McNally Map Company.
121. National Oceanic and Atmospheric Administration.
122. Anna Carpenter
123. From, *Subterranean Worlds 100,000 Years of Dragons, Dwarfs, the Dead Lost Races & UFO's from Inside the Earth,* Loompanics Unlimited, Port Townsend, Wa.
127. Map, Rand McNally Map Company, inset, Anna Carpenter
128. Anna Carpenter
129. Rand McNally Map Company
131. Top, Anna Carpenter, bottom, Rand McNally Map Company
135. From, *Subterranean Worlds 100,000 Years of Dragons, Dwarfs, the Dead Lost Races & UFO's From Inside the Earth,* Loompanics Unlimited, Port Townsend, Wa.

137. Top, Anna Carpenter, bottom, Rand McNally Map Company
138. Anna Carpenter
139. Both, Rand McNally Map Company
140. Anna Carpenter
141. Rand McNally Map Company
143. Rand McNally Map Company
144. Rand McNally Map Company
145. Rand McNally Map Company
147. National Oceanic and Atmospheric Administration.
153. Top, Anna Carpenter, bottom, Rand McNally Map Company
154. Anna Carpenter
157. Rand McNally Map Company
158. Rand McNally Map Company
160. Photo by author.
161. Anna Carpenter
164. Top, Anna Carpenter, bottom, Rand McNally Map Company
165. Top, Anna Carpenter, bottom, Rand McNally Map Company
166. NASA
167. Anna Carpenter
168. Top, Anna Carpenter, bottom, Rand McNally Map Company
169. Top, Anna Carpenter, bottom, Rand McNally Map Company
170. Top, Anna Carpenter, bottom, Rand McNally Map Company
171. Top, Anna Carpenter, bottom, Rand McNally Map Company
172. Top left, Rand McNally Map Company, top right, Anna Carpenter, bottom, author.
173. Top right, Anna Carpenter, bottom, author.
174. Top left & bottom, author, top right, Anna Carpenter
175. Photo by author
176. Photo by author
177. Both drawings by Piazzi Smyth
178. Top right, Anna Carpenter, bottom, U.S. Central Intelligence Agency.
179. Top left, Rand McNally Map Company, top right, Anna Carpenter, bottom, U.S. Central Intelligence Agency.
180. Top right, Anna Carpenter, bottom, U.S. Central Intelligence Agency.
181. Top left, Rand McNally Map Company, top right and bottom, Anna Carpenter.
182. Rand McNally Map Company
187. Map, Rand McNally Map Company, inset, Anna Carpenter
200. Anna Carpenter
203. Anna Carpenter

INDEX

Job 38:16

"Hast thou entered into the springs of the sea? or hast thou walked in the search of the depth?"